PLANT SCIENCE RESEARCH AND PRACTICES

MICROPROPAGATION OF ORNAMENTAL PLANTS

RESEARCH AND PRACTICE

Plant Science Research and Practices

Additional books and e-books in this series can be found on Nova's website under the Series tab.

PLANT SCIENCE RESEARCH AND PRACTICES

MICROPROPAGATION OF ORNAMENTAL PLANTS

RESEARCH AND PRACTICE

TAOUFIK BETTAIEB
EDITOR

nova
science publishers
New York

Library of Congress Cataloging-in-Publication Data

Names: Bettaieb, Taoufik, editor.
Title: Micropropagation of ornamental plants : research and practice / editor: Taoufik Bettaieb.
Description: Hauppauge, New York : Nova Science Publishers, Inc., 2018. | Series: Plant science research and practices | Includes index.
Identifiers: LCCN 2018044665 (print) | LCCN 2018047613 (ebook) | ISBN 9781536145427 (ebook) | ISBN 9781536145410 (hardcover)
Subjects: LCSH: Plants, Ornamental--Micropropagation. | Plant micropropagation.
Classification: LCC SB406.7 (ebook) | LCC SB406.7 .M53 2018 (print) | DDC 635.9/153--dc23
LC record available at https://lccn.loc.gov/2018044665

Published by Nova Science Publishers, Inc. † New York

CONTENTS

PREFACE

The book, *Micropropagation of Ornamental Plants: Research and Practice*, provides up-to-date reviews and protocols of the applications of micropropagation. It aims to be a useful resource, for experienced and knowledgeable professionals as well as an aid to newcomers, in order to become acquainted with the theory and practice of micropropagation applied to a wide array of ornamental plants including herbaceous, woody and monocots ornamental plants.

A total of 9 chapters are included, split into four sections.

Section I of this book, recalls fundamentals, current practices of plants micropropagation and traces briefly the history of this technology and the origin of the main ways it has opened (Chapter 1).

Section II details *in vitro* propagation of numerous herbaceous ornamental plants including *Aloe, Asparagus, Allium, Otochilus, Phytamax* using recent micropropagation techniques (chapter 2). It presents also a board group of new plant tissue culture tools that improve shoot multiplication rate of *Phalaenopsis* (chapter 3) and details a developed protocol for large scale production of *Lilium longiflurum* L. (chapter 4). The final chapter of this section, chapter 5, describes a cost and time-effective *in vitro* propagation scheme for cormel production of *Gladiolus* cultivars.

Section III presents overviews of *in vitro* propagation protocols of tow woody ornamental plants via organogenesis and somatic embryogenesis. In chapter 6 protocols for micropropagation of *Jacaranda mimosaefolia* D. Don by shoot-tip culture, alginate-encapsulation and germplasm conservation approach are described. Chapter 7 describes the initiation of embryogenic cultures of Ryukyumatsu (*Pinus luchuensis* Mayr.), from seed explants production of somatic embryos, regeneration of plants, and protoplast isolation from embryogenic tissues.

Finally, Section IV focused on advances micropropagation of ornamental monocots. Chapter 8 reviewed in great depth principles, strategies and challenges of somatic embryogenesis as a reliable way for propagation of ornamental palm trees. Chapter 9 contains a step-wise protocol for micropropagation of Bird of paradise (*Strelizia reginae* Ait.) and emphasizes on the most effective media to control oxidative browning of explants.

Collectively, these review chapters and the profusely illustrated protocols highlight the formidable power and versatility of micro-propagation applied for several economically important horticultural species.

At the end, I would like to express my sincere appreciation and gratitude to all the authors who have contributed to this book.

Taoufik Bettaieb

PART I: THEORETICAL BASIS OF *IN VITRO* PROPAGATION

In: Micropropagation of Ornamental Plants ISBN: 978-1-53614-541-0
Editor: Taoufik Bettaieb

Chapter 1

In vitro Multiplication of Plant Tissue Culture: Foundations and Practices

Sihem Soufi, Sarra Arbaoui and Taoufik Bettaieb*
Laboratory of Horticultural Sciences,
National Agronomic Institute of Tunisia,
University of Carthage, Tunisia

Abstract

In vitro multiplication (or micropropagation) as an alternative means of propagation allowed the production of a large scale number of genetically identical plants within a relatively short span of time. This process had made ornamental species multiplication possible where other propagation techniques were not effective to apply. This review briefly traces the history of this technology and the origin of the main ways it has opened and presents an overview on the *in vitro* propagation via organogenesis and somatic embryogenesis. The influences of different cultural conditions in the establishment of aseptic culture, multiplication,

* Corresponding Author Email: tbettaieb@yahoo.fr.

rooting and acclimatization phases on the growth of tissue cultured plant are also discussed.

Keywords: micropropagation, organogenesis, somatic embryogenesis

INTRODUCTION

Higher plant bodies have structurally well-differentiated organs with well-defined functions such as stems, leaves, roots and flowers which are vegetatively multiplied in addition to seeds. This way of multiplication allows the reproduction of individuals from somatic cells with 2n chromosomes without meiosis. Apart from few exceptions, the obtained plants are identical to the mother plant. Reproductive capacity of higher plants is dependent on the totipotency of plant cells. These cells are capable to regenerate an entire plant under appropriate nutritional, hormonal and environmental conditions. Currently, it is possible to vegetatively reproduce the plant from an organ, a segment of organ or even from a single plant cell.

Individual regeneration begins by a dedifferentiation followed by a differentiation at cellular level during which a loss and resurgence of specific cell function occur. These phenomena characterized plant cells that may potentially turn over to a juvenile stage, divided and specialized again into meristems that will be the origin of new organs. Cell morphology can be drastically changed during differentiation stage but genetic material remains the same with some exemptions. Dedifferentiation and differentiation are extremely rare in animal cell. The most blatant example in that respect is that of the triton tail where cells of the stump can reform all the tail tissues (Marcel and Vriz, 2003).

Vegetatively produced progeny called clones are genetically identical. These progeny are originated from a unique plant known as plant mother. Relying on plant cell totipotency, clones are broadly obtained spontaneously by asexual reproduction either by emission of suckers, on stolons or bulblets in the case of bulbous plant. Clones are also obtained by means of induced multiplication in nursery or in field by either cutting, grafting, layering or by *in vitro* propagation at laboratory level which allow the large scale production of healthy and homogenous plants. In this chapter we will discuss different aspects of micropropagation techniques namely, organogenesis and somatic embryogenesis and various factors affecting the micropropagation of plants.

HISTORY OF PLANT TISSUE CULTURE

The first biologist to enter the history of *in vitro* plant tissue culture is Gottlieb Haberlandt (1902) who stated the notion of plant cell totipotency. This property demonstrates that plant cells, taken from any plant organ, have the ability to regenerate a complete individual identical to the parent plant. This phenomenon is based on the ability of plant cell to dedifferentiate and return to a state of non specialization and then differentiate to give again a specialized cell. Therefore, the discovery of this plant cell property has formed the basis for plant tissue *in vitro* culture and has opened the way for several applications of this biotechnology in agriculture.

In 1926, Went discovered the auxins. However, their chemical structure has not been definitively identified till 1931 by Kogl and Haagen-smit. In the same year, Kurosawa identified gibberellin in a fungi *Gibberella fujikuroi* that excessively lengthens rice stems. Yabuta (1935) isolated and purified this substance. The chemical structure of gibberellic acid (GA3) was not determined in 1954.

The discovery of auxin allowed White (1934) to succeed the first tomato root culture in a medium containing water, mineral salts, yeast extract, sugar and auxin.

Gautheret (1939) obtained calli corresponding to cluster of dedifferentiated cells from carrot plant tissues. This discovery opens the way to several applications of *in vitro* culture.

Viral sanitation of plants was achieved for the first time thanks to the study of Ball (1946) as well as that of Morel and Martin (1952) who used as explants apical meristem of ornamental plants. Indeed, Ball regenerated vitroplants of Lupine and Capucine. Six years later, Morel and Martin demonstrated in Dahlia, that the apical meristem cultivated *in vitro* allowed the sanitation of plants affected by viruses.

In addition to auxins and gibberellins, Miller et al. (1955) discovered that other plant hormones, which are cytokinins, induce cell division and stem formation (caulogenesis) in tobacco tissue cultures. Moreover, under the influence of auxin and cytokinin, Skoog and Miller (1957) regenerated roots and stems from calli and showed that these growth hormones are essential for the plant development.

The first embryos obtained from somatic cells were obtained by Steward et al. (1958) from carrot root culture. This fact allowed to highlight the concept of cell totipotency and to open the way for the development of a new vegetative propagation method currently used at a large scale, namely, somatic embryogenesis.

In 1962, Murashige and Skoog developed the most widely used *in vitro* culture medium to date by studying its effect on the vegetative propagation of tobacco tissue cultures.

Producing haploid plants began by Guha and Mahereshwri (1964) who produced haploid plants from *Datura innoxia* Mill. from another culture. This discovery was the start of haplodiploidization which constitutes, at this time, a new way of genetic improvement.

Somatic hybridization, another biotechnological pathway for plant breeding, was known by Takebe et al. (1971) after regeneration of whole plants of *Nicotiana tabacuma* by protoplast fusion.

The first transgenic plants were obtained in 1983 in Belgium and the United States, the plant species were tobacco and petunias resistant to kanamycin antibiotics (Bevan et al. 1983; Fraley et al. 1983; Herrera-Estrella et al. 1983) and sunflowers expressing a bean protein (Murai et al. 1983).

IN VITRO PLANT PROPAGATION

Plant tissue culture is recognized as a useful tool that has multiple applications. Besides *in vitro* propagation or micropropagation, that is broadly used for virus cleansing, which uses meristems as explants. Several biotechnological applications of plant tissue culture have also been reported to be involved mainly in genetic improvement of plant such as haplodiploidization through male or female gamete cultures, somatic hybridization by protoplast fusion, vitrovariation by inducing mutations on *in vitro* tissues, embryo rescue for distant hybridization and obtaining genetically modified plants. The following section will only focus on micropropagation.

Routes of *In vitro* Plant Propagation

Micropropagation allows the production of a large number of genetically identical plants within a relatively short span of time and space, and that independently of season. This process had made rare species multiplication possible where other propagation techniques were not effective to apply. This has led to a large diversity of commercialized plants. Micropropagation techniques essentially use two pathways, namely, organogenesis and somatic embryogenesis. However, some particular models are adopted to simplify the commonly used techniques.

In vitro Multiplication by Organogenesis

The process of micropropagation via organogenesis with the formation of axillary budding, neoformation of adventitious buds or rhizogenesis involves 4 steps:

Step 1: Establishment of aseptic cultures
Step 2: *In vitro* shoots multiplication

Step 3: Rooting
Step 4: Acclimatization of vitroplants

The selection or development of the culture media is a prominent and required step to ensure tissue culture success. Culture media used are basically composed of distilled water, mineral salts, vitamins, sugar, and growth regulators. Culture media are generally solidified with agar. A number of media have been formulated, but Murashige and Skoog (MS) (1962) formulation is the most widely used. A better knowledge of the roles of plant growth regulators, according to the cell culture purpose, is also required to ensure the success of the *in vitro* multiplication program.

The most commonly *in vitro* used auxins are Indole-3-acetic acid (IAA), Indole-3-Butyric Acid (IBA), 1-Naphthaleneacetic acid (NAA) and 2,4-Dichlorophenoxyacetic acid (2,4-D).... These auxins are required by most plant cells for division and elongation, for differentiation of vascular tissues and for root induction. At high concentrations, most synthetic auxins are phytotoxic and can likewise work as herbicides (Sachs and Thimann, 1967).

Cytokinins like auxins are also known as a key regulator for various aspects of plant growth and development. The main *in vitro* used cytokinin are 6-benzylaminopurine (BAP), 6- furfuryaminopurine (kinetin), 4-hydroxy-3-méthyl-trans2-buténylaminopurine (zeatine), N6(2-isopentyl) adenin (2iP), 1-phenyl-3-(1,2,3, thiadiazol-5-y1)(TDZ).... Cytokinins promote cell division, shoot proliferation, and shoot morphogenesis (Sachs and Thimann, 1967).

GA3 is frequently added to plant tissue culture media and it promotes stem elongation by internodes elongation. Paradoxically, a number of studies have reported that for multiplication of certain ornamental species, a substance called anti–gibberellins has been used, such as coumarin and ancymidol in order to reduce internodes length and to obtain an interesting apical meristem clumps. Auxins and cytokinin are essential for plant growth and development, however, optimal ratio of Auxins/cytokinins is genotype dependent since their endogenous hormone content and apical degree of dominance are different.

During the first stage, aseptic technique is absolutely necessary for a successful establishment and maintenance of microcuttings or plant fragments cultured *in vitro*. During this stage different disinfectants are effectively used such as sodium hypochlorite, alcohol, calcium hypochlorite, etc. Surface sterilized explants are then inoculated on basal media supplemented with auxins and cytokinin following a specific ratio which depends on the genotypes. When apical meristem is used, GA3 can be used to promote the lengthening of the apical region of the explants. Depending on species, culture establishment under aseptic conditions lasts about 4 weeks for herbaceous species and even 2 to 3 months for slow-growing species. For these species culture medium should be renewed monthly. This phase is considered complete when contamination-free explants exhibit a positive morphogenetic response with a vegetative resumption.

During the multiplication phase, newly formed shoots are divided into different cuttings and are repeatedly subcultured onto new culturing media under aseptic conditions. Consecutive subculture enhances shoots proliferation and supplies shoots for the subsequent propagation phases but also maintains plant material stock. The subculturing should be limited in number in order to avoid mutations that may occur spontaneously during subculture. For slowly growing species, subculture lasts 4 to 6 weeks. Multiplication rates depend also on species. They range from, nearly, 2 for recalcitrant species to 30 shoots in easily multiplied species.

Proliferation is the result of the separation of preformed axillary buds, neoformed buds, or the division into microcuttings of nodes of elongated shoots. The auxin / cytokinin ratio is, logically, less than 1 in the first two cases and greater than 1 in the third case with even the possibility of not adding cytokinins. The subsequently *in vitro* developed buds could be pre-existing buds, terminal or axillary, adventitious neoformed on the explants or on a callus. Cytological studies have reported that *in vitro* neoformed buds from an organ fragment containing various tissues. These studies unraveled that the caulogenesis ability is highly dependent on the nature of the plant organ from which the explants was derived. It is maximal for cambial tissues, relatively high for phloem and xylem tissues, extremely low to zero for cortical or medullary parenchyma (Margara, 1989).

The third step involved in the process of micropropagation via organogenesis is rooting phase which is essentially designed for the establishment of fully developed plantlets. During the rooting phase, well developed vitroshoots of 2–5 cm high are transferred to rooting media supplemented with auxins. The remaining small vitroshoots are transferred to elongation hormone free media before rooting treatments. Rooting phase is almost the last *in vitro* stage before transferring the plantlets to *ex vitro* conditions. Thus, it is necessary to accustom vitroplants to autotrophic conditions and that by reducing mineral salts and sugar concentrations in rooting media.

According to Boxus et al. (1995), rhizogenesis is a complex and remarkable process that occurs in different phases. This complex phenomenon begins with cell dedifferentiation which results in the formation of callus cells that differentiate to form primordial root that develop into young roots. Generally, it is the cells of the secondary meristem of the libero-woody generating base or cambium which give the callus. The meristematic cells differentiate and organize to give birth to a new root. However, some neighboring cells are the origin of the roots. Boxus et al. (1995) on vine model also notes that primary phloem cells react first. The non-lignified cells of secondary origin and the cortical parenchyma, or even the marrow can also react by forming a more or less important scar tissue.

The final step in a successful micropropagation system is acclimatization to the *ex vitro* environment (Preece and Sutter, 1991). It has been noted that vitroplants may exhibit a reduced differentiation of palisade cells and also show large inter-cellular spaces and a lack of the protective mechanisms, namely waxy cuticles, stomatal regulation or leaf hairs, and which may lead to a rapid desiccation or wiltness and may even cause death of vitroplants as a result of change in environment, unless substantial precautions are taken to accommodate them (Nath Hazarika et al. 2015). Vitroplants are delicate and fragile so they should be placed in a greenhouse under a specific environment where humidity should be close to saturation, at least the first two weeks, and ambient and substrate temperatures should range from 18 to 25 °C. Direct exposure to sunlight should be avoided during this phase which requires the use of 50 to 70% shading cloth or the painting of the greenhouses cover.

Somatic embryogenesis is a remarkable process used with ease to multiply ornamental plants. It consists in the induction of somatic embryos from cells of explants cultured *in vitro*. As one of the most reliable techniques of biology, somatic embryogenesis allows somatic cells (nonzygotic plant cells), including haploid cells, to differentiate into somatic embryos and to regenerate fertile plants (Rose et al. 2010). Somatic embryogenesis is also considered as an asexual process that relies on the utilization of the intrinsic totipotency of non- germ cells, which involves dedifferentiation of a nonzygotic cell and subsequent redifferentiation to regenerate full plants (Sahrawat et al. 2003).

MICROPROPAGATION VIA SOMATIC EMBRYOGENESIS

Somatic embryogenesis is respectively referred to as direct or indirect according to the presence or absence of a callogenesis phase (Rout et al. 2006). Mathew and Philip (2003) have reported that in direct somatic embryogenesis, embryos are directly produced from pre-embryongenic determined cells without an intervening callus formation. Explants for direct embryogenesis include microspores, ovules, embryos and seedlings. Regarding the indirect somatic embryogenesis, somatic tissues are dedifferentiated into an unorganized mass of dividing cells, called calli which develops somatic embryos. Generally, direct somatic embryogenesis is rare in comparison with indirect somatic embryogenesis. It is worth noting that undifferentiated cells could be the source of a genetic instability. Therefore, in order to maintain conformity and quality of cloned plants, it appears that somatic embryogenesis should be limited to species with a very stable genome.

Somatic embryos, which are bipolar structures, arise from non-sexual plant tissues, and have no vascular connection with the maternal tissue of the explants (Haccius, 1978). As described by Pâques and Bercetche (1995), the somatic embryogenesis process involves several steps that include initiation and proliferation of embryogenic cell lines followed by maturation and germination of somatic embryos, up to the stage where somatic embryo plants (emblings) are transferred to *ex vitro* conditions. However, development of somatic embryos is highly dependent upon the type and concentration of plant growth regulators used (Poonam et al. 2004). As it is previously highlighted, intervening callus formation is an essential step for somatic embryos formation. This callus is generated from *in vitro* cultured plant organs (e.g., fragments of stem, leaf, flower). Usually, 2, 4-D or ANA is found to be effective for callus induction and for callus growth. However, hormonal composition is genotype dependent. After isolation, primary calli are maintained by sub culturing on the same medium. Once globular structures that evoke proembryogenic, masses appear on callus, this latter should be dissociated and inoculated on liquid multiplication or expression media which is generally devoid of auxin. Callus multiplies and forms an embryogenic tissue. The formed embyogenic tissue is established under an orbital shaking cell suspension. These multicellular aggregates are organized later into somatic embryos.

The obtained somatic embryos are encapsulated without altering their viability to produce synthetic seeds. The seed consists of somatic embryos and artificial endosperms. The somatic embryos coated by encapsulation matrix of sodium alginate, polyethyleneimine (Kersulec et al. 1993) or chitosane (Tay et al. 1993) and subsequently form the artificial endosperm. The condition of an artificial endosperm could provide nutrients and growth regulators that are essential for plantlet development from encapsulated parts (Nieves et al. 1998). It is therefore possible to preserve these seeds by cryopreservation in liquid nitrogen.

Simplified Models and Automation of Micropropagation

The transfer of somatic embryos into a growth medium deprived of hormone or enriched with active charcoal allows their germination and their conversion into rooted plants which will be acclimated in a greenhouse under the same conditions required for the acclimation of vitroplants resulting from the multiplication by organogenesis (Gaspar, 1988).

Transfer of shoots, plantlets or somatic embryos produced *in vitro* from developmental stage to another incurs a relatively high cost and labour input. Automation of either one or all various stages of organogenesis or embryogenesis is considered as a reliable way that reduces time labour input and the cost of plant or the somatic seedling (Aitken-Christie et al. 1995). In fact, a number of plant tissue culture aspects have been already automated which include preparation media, control of the culture environment, and computer management in greenhouses (Aitken-Christie 1991). The developed automation method depends on growth and multiplication habits of species, the developmental stage.

Multiplication of meristematic clusters is among the developed labour-saving systems that have been applied in microprpagation of Gladiolus (Ziv, 1990) d'Eucalyptus (Arezki et al. 2000). In fact, these authors add anti gibberellins to culture media which reduce the elongation of the internodes of cultivated plants. This is the origin of the formation of aggregates bearing very close meristems. The blind cutting of this material, simple and fast, allows in all cases to obtain multiplication units bearing meristems. This technique allows a huge gain in labor costs. Aitken-Christie (1991) has developed a multiplication system that is based on computer-guided visualization and automatic cutting, that identifies the orientation of the shoots to mark the site of the cut. Prototypes of existing robots are adapted to ports of species that form long shoots or species that form aggregates of meristems.

Inputs of Biotechnology to the *In vitro* Propagation of Ornamental Plants

The first outstanding applications of plant tissue techniques for ornamental species are those of Morel and Martin (1952) who obtained free viral diseases Dahlia plants by meristem culture. Then, Morel (1960) showed the possibility of rapid cultures with a high rate of cymbidium micropropagation. Indeed, the apical meristem of this ornamental orchid produces a protocorm capable of regenerating other protocorms. When separated each month and cultivated *in vitro* in a medium supplemented with cytokinin, these protocorms produce several identical structures.

Micropropagtion of numerous numbers of ornamental plants are associated with sanitation against viral diseases. For ornamental plants, viruses are vegetatively spread to offspring. The accumulation of several types of viruses during successive multiplications leads to the degeneration of the obtained plants which affects their aesthetic and/or floriferous qualities. The breaking in tulips is due to one or more viruses occurring as strains of varying severity. In most plants, stunting is also observed. Regeneration and *in vitro* propagation of ornamental plants from apical meristems is considered as an unavoidable way to obtain healthy and disease-free plant clones.

Besides apical meristem, other types of explants were used for massive multiplication of ornamental plants. In *Gerbera jamesonii*, the starting explant used for its *in vitro* propagation was the flower bud. In most indoor plants, such as *Ficus*, *Schefflera*, *Aralia*, etc., microcuttings were most often used.

As a result to these researches, the previously mentioned micropropagation protocols have been mastered on several numbers of ornamental plants. Thenceforward, industrial multiplication of plants by *in vitro* methods has been adopted since the 1970s by several laboratories around the world. The multiplied ornamentals include orchids, bulbous ornamentals (*Lilium, Hypeastrum, Narcissus ...*), houseplants (*Nephrolepis,*

Ficus, Anthurium ...) and a limited number of ornamental trees, many of which are recalcitrant to this mode of propagation.

Since the obtaining of the first *in vitro* somatic embryos on carrot tissue by Steward et al. (1958), several micropropagation protocols of ornamental plants have been developed using this way. Thus, the number of ornamental plants multiplied by somatic embryogenesis continues to increase. The first multiplied plants are: *Alstromeria, Anthurium andreanum, Camellia, Freesia, Lisianthus, Ranunculus asiaticus* ... This technique, which allows more multiplication than micropropagation by organogenesis, is an effective tool in this field when the genetic stability of the multiplied genotype is ensured. If not, it may be an interesting method of selecting new mutants.

References

Aitken–Christie J. 1991. Automation. In: P. C. Debergh, and R. J. Zimmerman (Eds.), *Micropropagation: Technology and Application* (pp. 363–388). Kluwer Academic Publishers, Dordrecht: The Netherlands.

Arezki, O., Boxus, P., Kevers, C., and Gaspar. (2000). Hormonal control of proliferation in meristematic agglomerates of *Eucalyptus camaldulensis* Dehn. *In vitro Cellular and Developmental Biology Plant, 36,* 398–40.

Ball, E. (1946). Development in sterile culture of stem tips and subjacent regions of *Tropaeolum majus* and *Lupinus albus* L. *American Journal of Botany, 33,* 301-318.

Bevan, M. W., Flavell, R. B., and Chilton, M. D. (1983). A chimaeric antibiotic resistance gene as a selectable marker for plant cell transformation. *Nature, 304*, 184-118.

Boxus, P., Jemmali, A., and Piéron, S. (1995). Vegetative propagation: Micropropagation. Chapter X. Stage II: Organogenesis or neoformation of buds. In: *Plant Biotechnology*. Demarly Y. & E. Picard (eds.). UNISA T, CNED/AUPELF-UREF. 63-78.

Conover, C. A., and Poole, R. T. (1984). Acclimatization of indoor foliage plants. *Horticultural Reviews 6,* 119-154.

Fari, M., (1987). Development of technical and biological basis of micropropagation toward automation and industrialization. In: *Electronics and management of Plants* (pp. 55-56). Moët Hennessy: Paris.

Fraley, R. T., Rogers, Horsch, S. G., R. B., Sanders, P. R., Flick, J. S., Adams, S. P., Bittner, M. L., Brand, L. A., Fink, C. L., Fry, J. S., Galluppi, G. R., Goldberg, S. B., Hoffmann, N. L., and Woo, S. C. (1983). Expression of bacterial genes in plant cells. *Proceedings of the National Academy of Sciences, USA* 80, 4803-4807.

Gaspar T. 1988. *In vitro* culture of higher plants. Chapter III. In: J. P. Zryd (Eds.), *Cultures of cells, tissues and plant organs. Theoretical and practical foundation* (pp. 31-49). Presses Polytechniques Romandes, Lausanne : Suisse.

Gautheret, R. J. 1939. On the possibility of achieving the indefinite culture carrot tuber tissues. *Weekly reports of the sessions of the Academy of Sciences, 208,* 118-120.

Guha, S., and Maheshwari, S. C. (1964). *In vitro* production of embryos from anthers of Datura. *Nature, 204,* 497.

Haberlandt, G. (1902). Culture experiments with isolated plant cells. Sitzungsber. *Akad. Vienna., Math. Cl.,* Dept., 1 (111), 69-92.

Haccius, B. (1978). Question of unicellular origin of non-zygote embryos in callus cultures. *Phytomorphology, 28,* 74-81.

Hazarika Budhindra Nath, Jaime A. da Silva Teixeira • Talukdar Akshay (2015) Effective Acclimatization of *In vitro* Cultured Plants: Methods, Physiology and Genetics. In: Teixeira da Silva, J. A. (Eds.), *Floriculture, Ornamental and Plant Biotechnology Advances and Topical Issues,* vol 2, (pp. 427-438). Global Science Books.

Herrera-Estrella, L., Depicker, Van Montagu, M., and Schell, J. (1983). Expression of chimaeric genes transferred into plant cells using a Ti-plasmid-derived vector. *Nature, 303,* 209-213.

Kogl, F., and Haagen-Smit, A. J. (1931). On the Chemistry of the Growth Substance K. Akad. Wetenschap. Amsterdam. *Proc. Sect. Sci., 34,*1411-1416.

Marcel, T., and Vriz, S. (2003). Vertebrate Appendage Regeneration: An Early Experimental Model for Studying Adult Stem Cells. *Medecine/Sciences, 19*(4), 65-71.

Margara, J. (1989). Bases of vegetative propagation. *Meristems and organogenesis.* I.N.R.A. Versailles. 262.

Mathew, M. M. and Philip, V. J. (2003). Somatic embryogenesis versus zygotic embryogenesis in *Ensete superbum. Plant Cell, Tissue and Organ Culture, 72,* 267-275.

Miller, C., Skoog, F., Von Saltza, M. H., andStrong, F. M. (1955). Kinetin, a cell division factor from desoxyribonucleic acid. *Journal of the American Chemical Society*, *77*, 1392.

Morel, G. 1960. Producing virus-fee cymbidium. *American Orchid Society Bulletin*, *29*, 495–497.

Morel, G. and C. Martin, 1952. Healing Dahlias with a virus disease. *Accounts of the Academy of Sciences*, *Paris, 235,* 1324-1325.

Murai, N., Sutton, D. W., Murray, M. G., Slightom, J. L., Merlo, D. J., Reichert, N. A., Sengupta-Gopalan, C., Stock, C. A., Barker, R. F., Kemp, J. D., and Hall, T. C. (1983). Phaseolin gene from bean is expressed after transfer to sunflower via tumor-inducing plasmid vectors. *Science, 222,* 476-482.

Murashige, T., and Skoog, F. (1962). A revised medium for rapid growth and bioassays with tobacco tissue cultures. *Physiologia Plantrum*, *15*, 473–497.

Pâques, M., and Bercetche, J. (1995). Vegetative propagation: somatic embryogenesis. In: Y. Demarly, and E. Picard (Eds.). *Plants Biotechnologies* (pp. 117-179). UNISA T, CNED/AUPELF-UREF.

Poonam, B., N. P. Bhatia and A. Nanjappa, (2004). Somatic embryogenesis in the nickel hyperaccumulating shrub, *Hybanthus floribundus* (liumdl.). F. Muell. *Plant Tissue Culture, 14,* 1-7.

Preece J. E., and Sutter E. G. (1991). Acclimatization of micropropagated plants to the greenhouse and field, In: P. C. Debergh, and R. H.

Zimmerman (Eds.), *Micropropagation. Technology and Application* (pp. 71–93). Kluwer: Dordrecht.

Rose, R. J., Mantiri, F. R., Kurdyukov, S., Chen, S. K., Wang, X. D., Nolan, K. E. and Sheahan, M. B. (2010). Developmental biology of somatic embryogenesis. In: EC. Pua and M. R. Davey (Eds.), *Plant developmental biology–biotechnological perspectives: Volume 2* (pp. 3-26) .Springer-Verlag Berlin Heidelberg.

Rout, G. R., Mohapatra, A. and Jain, S. M. (2006). Tissue culture of ornamental pot plants: a critical review on present scenario and future prospects. *Biotechnology Advances, 24*, 531-560.

Sachs T., and K. V. Thimann. 1967. The Role of Auxins and Cytokinins in the Release of Buds from Dominance *American Journal of Botany, 54*(1),136-144.

Sahrawat, A. K., Becker, D., Lütticke, S. and Lörz (2003). Genetic improvement of wheat via alien gene transfer, an assessment. *Plant Science, 165,* 1147-1168.

Simonton, W. (1992). Issues in robotic system design for transplant production systems. In: K. Kurata, and T. Kozai (Eds.), *Transplant production systems* (pp 103-115). Kluwer Academic Publishers: Dorderecht.

Simonton, W. (1992). Issues in robotic system design for transplant production systems; In: K. Kurata, and T. Kozai (Eds.), *Trasplant production systems* (pp. 103-115). Kluwer Academic Publishers: Dordrecht (The Netherlands).

Skoog, F., and Miller, C. O. (1957). Chemical regulation of growth and organ formation in plant tissue cultures *in vitro*. *Symposium of the Society of Experimental Biology*, *11*, 118–131.

Steward, F. C., Mapes, M. O., and Mears, K. (1958). Growth and organized development of cultured cells. II. Organization in cultures grown from freely suspended cells. *American Journal of Botany*, *45*, 705–708.

Takebe, I., Otsuki, Y., and Aoki, S. (1971). Infection of isolated tobacco mesophyll protoplasts by tobacco mosaic virus. *National Symposia. Scientific Research National Center, 193,* 503-511.

Went, F. W. (1926). On growth-accelerating substances in the coleoptile of *Avena sativa. Proceedings of the Section of Sciences, Royal Academy of Sciences in Amsterdam, 30,* 10-19.

White, P. R. (1934). Potentially unlimited growth of excised tomato root tips in a liquid medium. *Plant Physiology, 9,* 585–600.

Yabuta, T. (1935). Biochemistry of the "bakanae" fungus of rice. *Agriculture and Horticulture, 10,* 17–22.

Ziv, M., (1990). The effect of growth retardants on shoot proliferation and morphogenesis in liquid cultured *Gladiolus* plants. *Acta Horticulturae, 280,* 207–214.

PART II: *IN VITRO* PROPAGATION OF HERBACEOUS ORNAMENTAL PLANTS

In: Micropropagation of Ornamental Plants ISBN: 978-1-53614-541-0
Editor: Taoufik Bettaieb

Chapter 2

MICROPROPAGATION OF SOME HERBACEOUS ORNAMENTAL PLANTS

Satyesh Chandra Roy[1,*], Tui Ray[2], Prasenjit Saha[2] and Ashutosh Mukherjee[3]
[1]Department of Botany, University of Calcutta, Kolkata, India
[2]Department of Biochemistry,
University of Pennsylvania, Philadelphia, US
[3]Department of Botany, Vivekananda College,
Kolkata, West Bengal, India

ABSTRACT

Ornamental plants particularly the herbaceous ones are commonly the most important floral wealth of any country with the excellence of floral beauty, fragrance, colour and foliage leading to a great commercial and export value in the global market. With the increase in demand of these plants, an alternative method of multiplication called micropropropagation technique, has been developed and widely used to achieve commercial success. This is an overview on the *in vitro*

* Corresponding Author Email: scroyind@yahoo.com

propagation of horticultural plants using recent techniques to meet the demand of commercial production. Major challenges during micropropagation and the usefulness of checking genetic fidelity among the micropropagated plants utilizing cytological and molecular markers are discussed.

Keywords: micropropagation, culture systems, genetic fidelity, molecular markers

INTRODUCTION

Ornamental plants comprise plants that are broadly used for decorative purposes both for beautifications in outdoor areas like gardens, landscapes, road-side spaces etc. or for indoor purposes within office or rooms of the building. These are also displayed as cut flowers. All these plants have a tremendous value in drawing attention to the people and their Government for increasing the aesthetic value and in maintaining the global diversity in the formation of greener, pollution free and ecologically sustainable earth. With the development of urbanisation in modern days, more and more interests are growing in gardening and landscaping to enhance the beauty of buildings, communities and different cities to increase property values and human health leading to the demand in rapid propagation of ornamental plants.

Micropropagation, a well-researched area of biotechnology, has been exploited extensively for ornamental plants as an alternative to conventional method of cultivation. Although it is a true-to-type propagation of a selected genotype to increase the productivity for mass propagation of ornamental plants but this *in vitro* culture technique also has the ability to produce genetic variability to develop new and novel types through somaclones.

DIFFERENT MICROPROPAGATION SYSTEMS OR TECHNIQUES

Micropropagation comprises six important steps i) Initiation of cultures where explants may be shoot tip, meristem, bulbs, rhizome and other organs, ii) Shoot multiplication, iii) Somatic embryogenesis through the formation of embryogenic calli, iv) Formation of roots in plantlets produced *in vitro*, v) Acclimatization and vi) Transplantation of plants to field or pots.

There are some popular ornamental plants where different explants and micropropagation methods have been followed. For example, *in vitro* propagation of *Hydrangea* sp., popularly used for garden decoration and its cut flower branches are used in the commercial market (Ruffoni et al. 2013). The plant *Hydrangea* is famous for their colourful inflorescences especially of corymbs or panicles with a variety of colours like pink, blue, white and purple. Its flower has another unique characteristic as they change their colour with the changes in the pH of the soil. In acidic soil it is blue, in neutral soil it is cream and alkaline soils show pink or purple flowers. Of the several methods of micropropagation techniques used in the different species of *Hydrangea*, the method used by Ruffoni et al. (2013) is the best. They used MS media (Murashige and Skoog, 1962) for the initiation of explants and their multiplication was done with 0.25 mg/l of BA (Benzyl adenine) and the auxin NAA (0.5 mg/l) is used only for rooting without any BA at a light intensity of 35μmol/m^2/s (Ruffoni et al. 2013).

The micropropagation of *Gerbera jamesonii*, another ornamental herbaceous is widely accepted over vegetative multiplication to overcome long delay for commercial distribution. The micropropagation of *G. jamesonii* did not show any detectable morphological variation in regenerated plants (Figure 1) after culturing from young capitulum explants (Ray et al. 2005).

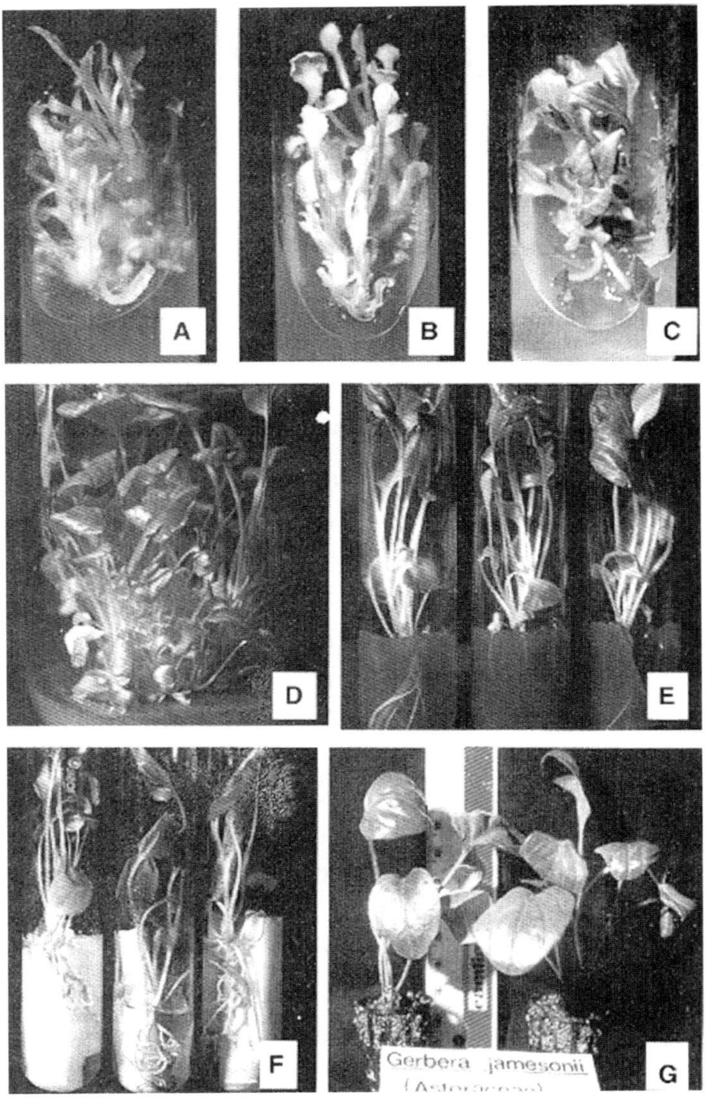

Figure 1. *In vitro* plant regeneration of *Gerbera jamesonii* through young capitulum explants. (A) Initial multiple shoot formation in MS medium supplemented with 7 mg/l BA and 0.1 mg/l IAA in dark. (B) Elongated multiple shoots in dark. (C) Green shoots after exposing to continuous light. (D) Numerous shoots with profuse leaves on ½ MS basal medium supplemented with 2 mg/l BA. (E) Rooting of regenerated shoots on ½ MS + 1.5 mg/l IAA + 0.5 mg/l IBA + activated charcoal. (F) Regenerated plants in liquid MS medium and (G) *G. jamesonii* plants established in soil (Ray et al. 2005).

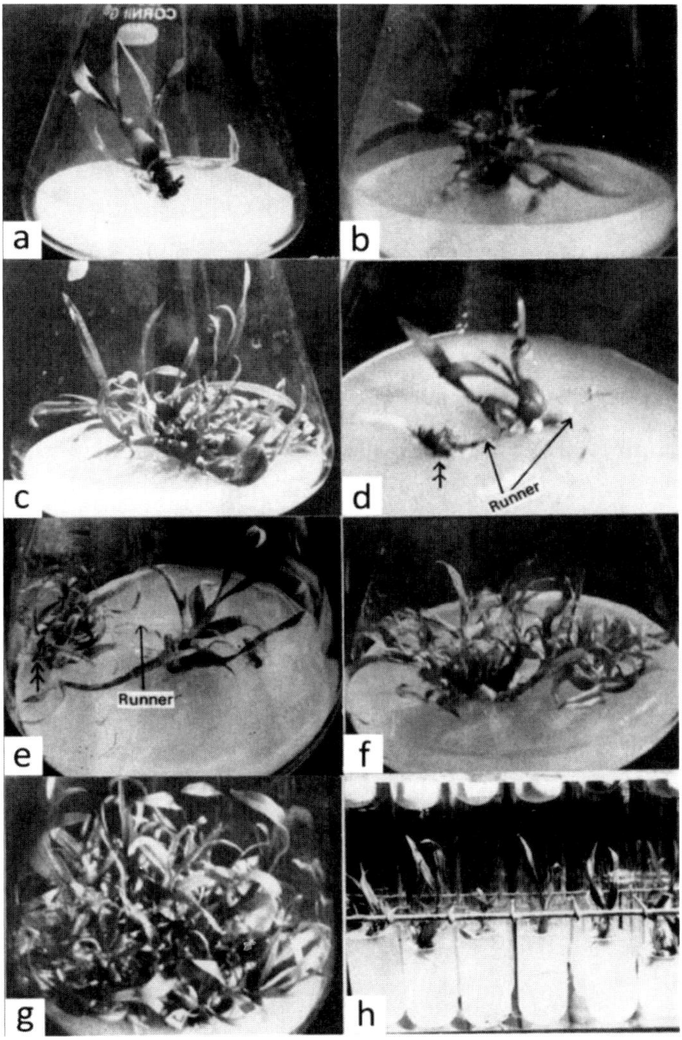

Figure 2. Formation of runners in *Otochilus alba*. (a) Shoot tip derived pseudobulbs after 5 weeks in MS medium. (b) Multiplication of pseudobulbs in MS media containing IAA (0.5 mg/1) and 2iP (2.0 mg/1). (c) Multiplication of pseudobulbs in MS media containing NAA (0.5 mg/1) and 2iP (2.0 mg/1). (d) Induction of two runners (arrow) in Phytamax media; and the formation of bulbous mass of tissue (double arrow). (e) Development and multiplication of pseudobulbs on the runner derived bulbous structures (arrow). (f) Profound multiplication of all pseudobulbs in Phytamax media after 35 days. (g) Innumerable pseudobulbs after 50 days of culture in Phytamax. (h) Induction of roots started in plantlets in MS media containing IBA (1.0 mg/1) after 7 days (Mukhopadhyay and Roy, 1994).

An efficient protocol for *in vitro* propagation of *Teucrium fruticans*, a valuable ornamental and medicinal plant was reported through axillary shoot proliferation (Frabetti et al. 2009). Node and shoot tip explants were used to develop plantlets through axillary shoot proliferation (Sivanesan et al. 2011).

Huernia hystrix, an endangered, highly demanded medicinal and ornamental succulent is efficiently propagated from stem explants cultured on MS media which showed root organogenic potential (Amoo et al. 2009). A perennial, monocotyledonous, ornamental plant, *Iris* sp., is micropropagated through meristems, bulbs or through splitting of rhizomes to avoid slow vegetative propagation (Jevremović and Radojević, 2006).

In an orchid *Otochilus alba*, pseudobulbs were produced from shoot apex in MS media containing 0.2 mg/l of IAA and NAA and the multiplication of pseudobulbs was done after transferring the culture to the media containing 2 mg/l 2iP and 0.5 mg/l NAA. When pseudobulbs were transferred to Phytamax Orchid multiplication medium (Sigma) runners were produced (Figure 2) from which several pseudobulbs were produced within 10 days leading to large number of plantlets (Mukhopadhyay and Roy, 1994).

Micropropagation of Some Economically Important Plants

Aloe vera

This species of *Aloe* and other species of this genus have ornamental, medicinal and cosmetic markets in the world. Conventionally these are vegetatively propagated and seed formation is very rare. Based on its slow growth and high demand in the market, the *in vitro* culture through axillary shoot from nodal segments of rhizomatous stem has been done for the multiplication of clones (Roy and Sarkar, 1991). Leaves of *A. vera* contain an important drug Aloin (Roy and Sarkar, 1991) and have more than 200 bioactive constituents (Waller et al. 1978). Meyer and Staden (1991) and

Roy and Sarkar (1991) used 2,4-D (0.02 mg/l) and Kinetin 1 mg/l for shoot initiation and multiplication (Figure 3a and 3b) while other authors used various combinations of hormones such as BA (0.2 mg/l) and 0.2 mg/l IBA (Aggarwal and Barna, 2004); BA (2.5 mg/l) and 4.0 mg/l Kinetin (Haque and Ghosh, 2013); BA (2 mg/l), IBA (0.5 mg/l) and 4.3 mg/l Adenine sulphate by Das et al. (2015) using MS medium in all cases. Again the multiplication of shoots were formed after reducing the strength of hormones in the media and rooting was done after culturing the plantlets either in hormone free media or by adding low concentration of auxin.

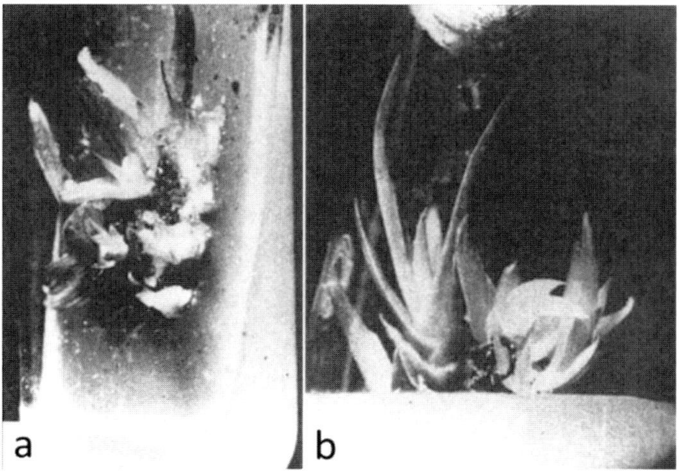

Figure 3. Micropropagation of *Aloe vera*. (a) Multiplication of shoots in media containing 2,4-D (0.02 mg/l) and Kn (1 mg/l) and (b) Formation of plantlets (Roy and Sarkar, 1980).

Asparagus

This plant has both ornamental and medicinal values. *Asparagus plumosis, A. densiflorus, A. virgatus, A. myriocladus and A. retrofractus* are ornamentals and *A. officinalis, A. racemosus, A. verticillatus, A. adscendens and A. curillus* are medicinal. Of these species, *Asparagus officinalis* is the most widely used in the garden as ornamentals as well as in medicine. Considering the importance of *A. officinalis* and slow growth in the conventional methods of cultivation, the technique of micro-propagation has been widely used by many laboratories from apical bud, axillary bud, nodal segments and through somatic embryo (Yang and Clore, 1973; Razdan, 2003; Štajner, 2013). Rapid *in vitro* propagation has been successful by using MS medium (Murashige and Skoog, 1962) containing NAA of 0.2 mg/l, Kinetin 0.2 mg/l and Ancymidol 0.1 mg/l (Štajner, 2013). The use of a growth retardant Ancymidol for rapid multiplication of shoots may be due to the suppression of apical growth as well as the alteration of carbohydrate metabolism in *A. officinalis* (Desjardins et al. 1987; Štajner, 2013). Ancymidol reduces time for production of transplantable plantlets from 20 to 8 weeks and regulates shoot growth and root formation of *Asparagus* (Štajner, 2013). He also increased the efficiency of acclimatization and formation of new *ex-vitro* grown shoots by trimming of shoots at the start of acclimatization in the greenhouse at 24 ± 2°C and relative humidity around 80%. In several plants including *Asparagus*, the efficiency of micropropagation has been found to be increased in liquid culture.

Mehrotra et al. (2007) showed that rate of shoot multiplication was increased to a great extent in liquid culture methods as shaking the culture in a shaker increases the axillary bud regeneration. In addition to the reduction of costs, the subculture of shoots or somatic embryos and root initiation in liquid media are very easy to do with minimum damage (Pati et al. 2005).

The process of somatic embryogenesis is another potential system for easy micropropagation through callus or directly from somatic cells in addition to a source of producing genetic variability to get new genotypes. Somatic embryogenesis through callus cultures have great potential for mass propagation as embryogenic calli can be easily transformed into cell suspensions leading to large number of somatic embryos in liquid cultures (Caligari and Shohet, 1993). The induction of callus is done in solidified MS media supplemented with 2 mg/l of 2,4-D and 3% Sucrose. Yang et al. (2000) transferred clusters of young calli in the liquid media containing 0.5 mg/l of 2,4-D and 1 mg/l of BA with 2% Sucrose and large number of somatic embryos (80%) were formed within 30 days. The somatic embryos were sieved and washed three times in MS liquid medium and were allowed to regenerate plantlets in half strength MS medium supplemented with 1 mg/l GA_3, 1 mg/l BA, 1% Sucrose and 0.8% agar (Yang et al. 2000). The micropropagated plants from somatic embryos and other explants were morphologically and cytologically similar. Haque and Ghosh (2013) have also found that there was no variation even after RAPD fingerprinting with the help of molecular markers in somaclones of *Aloe vera*.

Allium

Many members of the family Liliaceae and Amaryllidaceae are known as ornamental bulbous plants and are widely cultivated in ornamental gardening. Ottesen (2014) showed the importance of *Allium*s as indispensable ornamental plants for gardening in spring and summer display. Another reason for cultivating *Allium* species and hybrids as ornamentals are mostly due to their high commercial value in having additional medicinal and edible importance of Onions, Garlic, Leek, Chives and others. The most important ornamental *Allium*s are *A. afflatunense* (Persian), *A. giganteum*, *A. karateviense, A. neapolitanum, A. schoenoprassum* (Chives) etc. Some works on micropropagation of ornamental species like *A. giganteum* has been done by Šušek et al. (2002) showing direct organogenesis to multiple shoot structures from flower buds or ovaries without any callus intervention to avoid chromosome abnormalities. They have used this procedure as previous attempts of regeneration through callus caused chromosomal aberrations resulting in phenotypic instability in *A. fistulosum* (Lee and Ono, 1999), *A. tuberosum* (Roy, 1980, Do et al. 1999), *A. cyaneum* (Lee et al. 1998), *A. cepa* (Roy, 1980).

A very important medicinal and spice plant, Garlic (*A. sativum* L.) is conventionally propagated by daughter bulbs (cloves) and bulbils from the flower head. Another efficient protocol for *A. sativum* has been developed for somatic embryogenesis and up to 30% of the somatic embryos were converted into plants with shoots and roots after 8 weeks (Fereol et al. 2002). Generally micropropagation is not done through callus as the propagation via callus cultures causes many chromosomal abnormalities as reported by several authors (Roy, 1980; Lee et al. 1998, 1999). Micropropagation through meristem culture was also used for producing higher numbers of healthy plants free of viruses (Keller and Senula, 2013). As there is a chance of contamination in using underground bulbs as explants, Kozak and Stelmaszczuk (2013) did a two-step procedure in shaking bulb fragments in 2% Cl^- for 30 min and then soaking in 1% Cl^- for 15 minutes to get cultures free from any contamination in *A. aflatunense* and *A. karataviense. A. aflatunense* is one of the most decorative species for cut flowers industry while *A. karataviense* is suitable for planting in containers and in rock gardens (Kozak and Stelmaszczuk, 2013).

Another important ornamental species of *Allium* is *A. neapolitanum* Cirillo which is suitable for cultivation in flowerbeds, rock gardens as well as in containers. Stelmaszczuk and Kozak (2013) reported that the best multiplication of shoots in *A. neapolitanum*, were obtained on MS medium supplemented with 5 mg/l of BA and 0.1 mg/l of NAA during two subsequent subcultures. This method was successful as the survival rate of the plantlets in a peat:perlite (1:1 v/v) mixture after 4 weeks was 70% (Stelmaszczuk and Kozak, 2013).

Gantait et al. (2009) compared the field performance of micropropagated plants and conventionally propagated elephant garlic *A. ampeloprasum* L. where *in vitro* generated plants exhibited significantly more vigorous morphological growth.

A. altissimum Regel is used as an ornamental plant in landscape design. Poluboyarova and Novikova (2013) have developed a protocol of micropropagation using the uneven-aged flower buds from the inflorescence of *A. altissimum* and found that the best explants are the buds with the style length of 0.4-0.7 mm that are formed in the middle layer of the inflorescence. Wawrosch et al. (2001) described an appropriate method of micropropagation of *A. wallichii* Kunth where an average of 4.6

multiplication were possible on MS medium supplemented with 4.3 mg/l zeatin. The study reported that the plantlets were acclimatized to greenhouse conditions and transferred to the field successfully.

ORNAMENTAL FOLIAGE PLANTS

Ornamental foliage plants, popularly known as house plants, have very attractive leaves and can be grown indoors. These plants are widely distributed in the tropical and subtropical regions. The production of multicolour foliage plants has resulted in a dramatic increase in horticultural industry for the beautification in homes, offices, churches, hotels, malls and other places. Presently, most of the foliage plants are exported to Germany, US, Netherlands, UK, Italy, Japan and other countries (Ray et al. 2006a).

Cordyline terminalis

This plant is one of the most economically important indoor potted plants with beautiful and attractive multicoloured foliage. Since the growth of the plant is very slow, micropropagation technique has been widely used in foliage plants. The process of micropropagation generally forms true-to-type plants but may also produce somaclonal plants with new genotypes generating new desirable traits.

The explants used in the culture are the shoot tips in MS media containing different concentrations of IAA and cytokinins (Ray et al. 2013). The initiation of shoot buds was found to appear after 15-20 days in culture. Large numbers of shoots were produced in media containing BA (2 mg/l), IAA (0.1 mg/l) and 80 mg/l of Adenine sulphate (Figure 4). Plants were regenerated in half strength MS liquid media (Figure 4). Somaclonal variations were also found which was monitored by using isozyme markers. (Ray et al. 2006a).

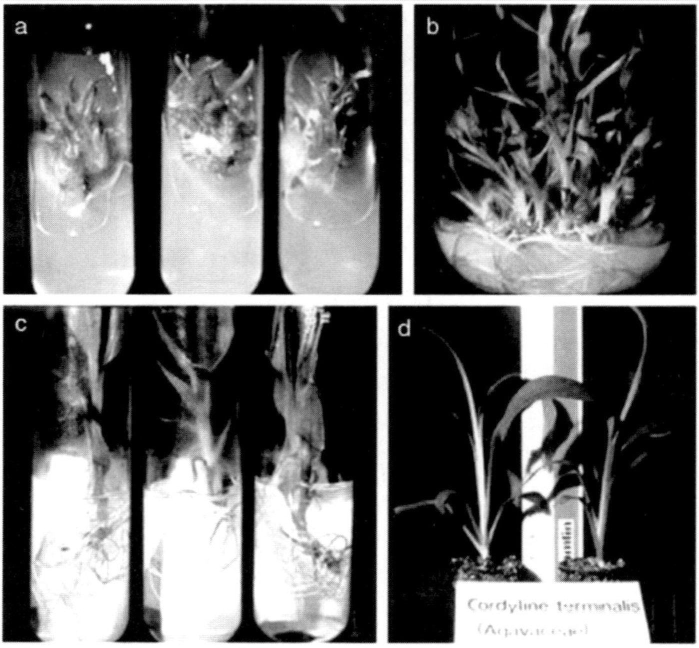

Figure 4. Different stages of micropropagation of *Cordyline terminalis* from shoot apex meristem. (a) Multiple shoot bud formation from shoot apex explants after 20 days in culture. (b) Numerous shoots with profuse leaves and roots on rooting medium. (c) Regenerated plants in half strength MS basal liquid medium. (d) Micropropagated plants, 20 days after transplanting in soil (Ray et al. 2006a).

***Dieffenbachia* sp.**

Dieffenbachia is an ornamental foliage plant with spotted and stripped with different beautiful colours. Conventional methods of propagation are very slow and are very susceptible to infection leading to loss of production of plants. *In vitro* propagation overcomes many problems like rapid multiplication, production of improved varieties and disease free plants (Govil and Gupta, 1997).

Shoot induction of this plant was done either from axillary bud or stem nodes on MS media containing 1.0 mg/l each of NAA and BA and multiplication of shoots were done after subculturing in media containing

5.0 mg/l BA and 0.5 mg/l of NAA (Elsheikh et al. 2013). Shen et al. (2008) showed somaclonal variation among regenerated plants through indirect organogenesis from three cultivars of *Dieffenbachia*.

Musa sp.

Many species of *Musa* are ornamentals with attractive foliage and flowers and some are edible. The efficient micropropagation protocols were in three important cultivars of *Musa* species of two genotypes like Robusta (AAA), Giant Governor (AAA) and Martaman (AAB) from shoot tip meristem (Ray et al. 2006b). Shoot initiation and multiplication was done in these cultivars of banana using B-SIM (Banana-Shoot Initiation and Multiplication) medium as reported by Mante and Tepper (1983). For further elongation of shoots these were cultured in half-strength MS medium supplemented with 5 mg/l BA, 100 mg/l ascorbic acid and 150 mg/l citric acid. For rooting, elongated shoots were transferred to a half-strength MS medium fortified with 3 mg/l IBA, 25 mg/l ascorbic acid and 50 mg/l citric acid (Figure 5). The well rooted plantlets were acclimatized in green house before they were transferred to field (Ray et al. 2006b).

SOMACLONAL VARIATION

Although the aim of plant production by micropropagation is to obtain the highest multiplication rate in lowest cost without any genetic change, it is now well recognized that the clonally produced plants are not always genetically uniform. Somaclonal variation is perhaps one of the greatest challenges to the extensive application of micropropagation and other tissue culture techniques to any plant (Peschke and Phillips, 1992).

A large number of reports indicate that genetic changes might occur during cell division (Larkin and Scowcroft, 1981) or differentiation (Vasil, 1985). It has been reported that genetic stability of tissue cultured clones mainly depends on genotype, the chimaeric nature of the explant used, cultural conditions, maintenance of cultures (Haisel et al. 2001) and number of subcultural passages (Chatterjee and Prakash, 1993).

Figure 5. Stages of micropropagation of banana from shoot-tip meristem explant. (a) Multiple shoot-initials formation from shoot-tip meristem explants after 30 days in B-SIM medium. (b) Cluster of shoots developed from shoot-initial. (c) Elongated multiple shoots with profuse leaves and roots on rooting medium. (d) Regenerated plant in pot containing soli-rite. (e) Micropropagated plants after transferring to soil (Ray et al. 2006b).

Somaclonal variations can produce new genotypes in ornamental plants with new floral patterns, colour variations or some new pattern in the foliage but it has some disadvantages also in not producing plants of same genotype.

Somaclonal variation can be minimized by using simple media with minimum supplement just for stimulation of existing axillary buds to grow-out rather than involving adventitious bud formation and also the formation of shoot buds without any callus intervention.

Decline of Growth during Culture

Occasionally, after a prolonged period of continuous subculture, cultured tissues may show decline in growth. This might be due to the fact that cells spontaneously acquire the capacity to synthesize cytokinin and will therefore able to grow on a medium without any added hormone. Habituation is a major limitation in commercial production systems resulting in the progressive decline in culture vigour and productivity with successive subculturing cycles.

The concept of habituation to growth regulators and the need to transfer cultures to hormone-free media periodically has now been recognized. Many current protocols involve shorter periods on hormone-containing media with an intervening period on hormone-free media. Less attention has been paid to progressive changes in other chemical, physical and biological components of the culture system over time. However, there is a need to look more closely at the other constituents of media, including the mineral supply.

Monitoring of Genetic Fidelity

Tissue culture can be worthwhile only if complete genetic fidelity of micropropagules is maintained. Genetic fidelity is the maintenance of genetic constitution of a particular clone throughout its growth span (Chatterjee and Prakash, 1993). Various markers including morphological, cytological, biochemical and molecular traits have been recommended to evaluate the genetic stability and clonal fidelity of the tissue culture grown plants. Monitoring genetic fidelity is crucial in mass propagation for commercial production of selected genotype since there are several reports of genetic changes during clonal propagation. Some key factors controlling genetic stability of tissue cultured clones are genotype, nature of the explant used, cultural conditions mainly medium composition, growth regulators and duration of cultures. Ray et al. (2006a) reported isozyme markers in studying genetic fidelity of somaclones of *C. terminalis* plants (Figure 6). The micropropagation protocol of *C. terminalis* showed no enzymatic polymorphism in somaclones indicating genetic stability. Probably using shoot apex meristem as an explant was one of the reasons for our isozyme results to be highly comparable in somaclones and their mother plant since the organized meristems do not undergo much genetic change during cell division (Larkin and Scowcroft, 1981).

However, few polymorphisms observed through isozyme studies, in Esterase and Super Oxide Dismutase, may be due to recurrent subculture for indefinite periods or the application of high concentrations of cytokinin in the culture medium under *in vitro* conditions (Ray et al. 2006a). Similar observation in esterase profiles after *in vitro* propagation had been reported by other researchers as well (Chatterjee and Prakash, 1993; Dutta et al. 2003). Numerous molecular markers have facilitated research on genetic variation at the DNA level. DNA fingerprinting has been utilized in detecting genetic stability of micropropagated plants in recent years. RAPD and ISSR markers revealed genotype dependent genetic stability during micropropagation of Banana (*Musa* sp.) cultivars, Robusta, Giant Governor, and Martaman from shoot tip meristem where Martaman variety (Figure 7) revealed complete genetic stability (Ray et al. 2006b).

The results observed in several varieties and cultivars of different species of *Allium* as well as two other related ornamental plants namely *Agapanthus africanus* and *Nothoscordum fragrans* (Figure 8) showed distinct genetic differences through the technique of DNA fingerprinting using RAPD and ISSR markers (Mukherjee et al. 2013). Thus, these techniques might be used as a good diagnostic tool for the assessment of genetic stability as well as variability of any cultivars and micropropagated plants.

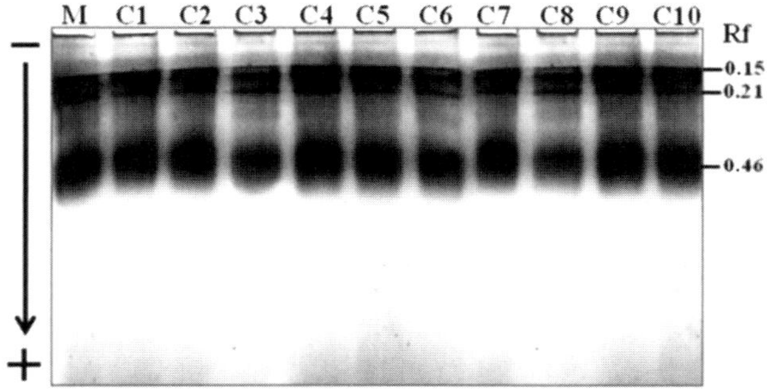

Figure 6. Zymogram of esterase isozyme of somaclones in *Cordyline terminalis*. Mother plant (M) and somaclones (C1–C10). Lanes C1–C5 and C6–C10 are *in vitro* plantlets (in culture vessel) and plants transferred to field after 10 and 15 subcultural passages, respectively (Ray et al. 2006a).

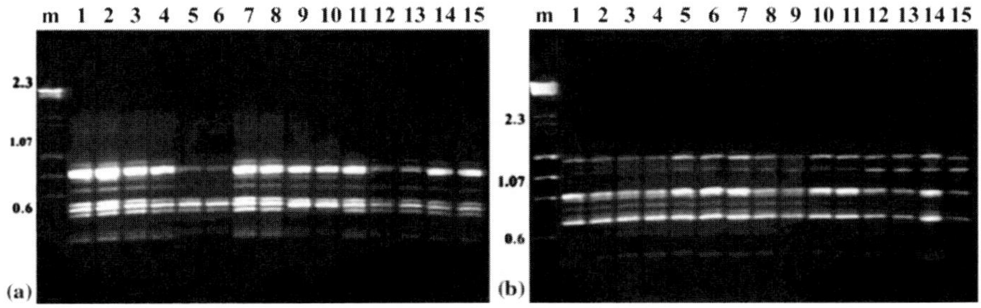

Figure 7. Gel electrophoresis pattern of RAPD and ISSR markers from micropropagated somaclones of Giant Governor cultivar of *Musa*. (a) RAPD profile obtained by the primer OPA-16. (b) ISSR profile obtained by the primer ISSCR-03. In both the figures, lane m λ DNA *Hind*III digest and Φx 174 DNA *Hae*III digest marker, 1 mother plant, 2–4 plantlets after ten subcultures, 5–15 plantlets after fifty subcultures (Ray et al. 2006b).

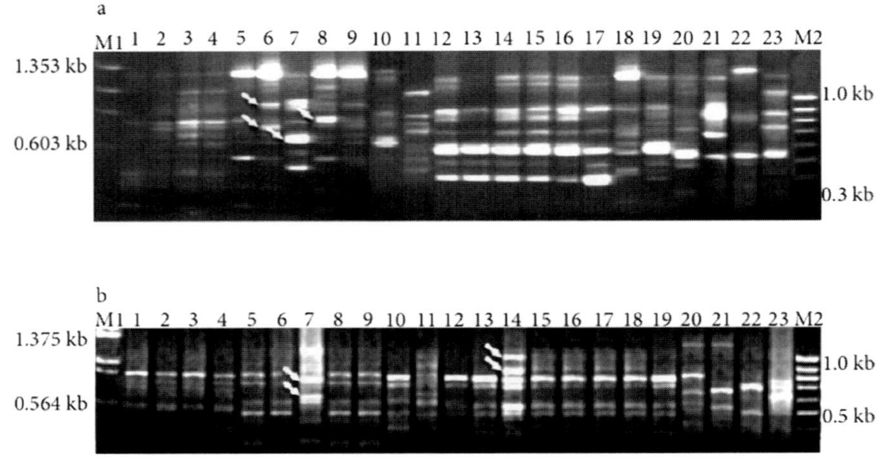

Figure 8. RAPD and ISSR patterns of different cultivars of *Allium*. a- RAPD profle obtained with primer OPG-2. M1: λDNA *Hind*III digest and ΦX 174 DNA *Hae*III digest marker; M2: 100 bp DNA ladder. Lane 1–4: *A. porrum* (1: Armor, 2: Alto, 3: Selecta, 4: Maridor); 5–9: *A. sativum* (5: Single clove, 6: Yamuna safed, 7: Bote lasun small, 8: Bote lasun large, 9: Agrifound parvati); 10: *A. cepa* var. *viviparum*; 11: Spring onion; 12: *A. cepa* var. *aggregatum*; 13–19: *A. cepa* var. *cepa* (13: Pusa white round, 14: Puna red, 15: Patna red, 16: Punjab selection, 17: Agrifound light red, 18: Pusa red, 19: Sukh sagar); 20: *A. stracheyi*; 21: *A. tuberosum*; 22: *Agapanthus africanus*; 23: *Nothoscordum fragrans*. Arrowheads indicate unique bands. b- ISSR profle obtained with primer UBC-807. M1: λ DNA *Eco*RI / *Hind*III double digest marker; M2: 100 bp DNA ladder. Sequence of materials is the same as in the RAPD profle. Arrowheads indicate unique bands. (Mukherjee et al. 2013).

References

Aggarwal, D., and Barna, K. S. (2004). Tissue culture propagation of elite plant of *Aloe vera* Linn. *Journal of Plant Biochemistry and Biotechnology, 13,* 77-79.

Amoo, S. O., Finnie, J. F., and Staden, J. V. (2009). *In vitro propagation* of *Huernia hystrix*: an endangered medicinal and ornamental succulent. *Plant Cell, Tissue and Organ Culture, 96*, 273-278.

Caligari, P. D. S., and Shohet, S. (1993). Variability in somatic embryos. *In* K. Redenbaugh (Eds.), *Synseeds: Applications of Synthetic Seeds to Crop Improvement* (pp. 163-174). CRC Press, Boca Raton, USA.

Chatterjee, G., and Prakash, J. (1993). Genetic stability in commercial tissue culture. *In* J. Prakash, R. L. M. Pierik (Eds.), *Plant Biotechnology: Commercial Prospects and Problems* (pp. 111-121). Oxford & I. B. H. Publishing Co., New Delhi: India.

Das, A., Haque, S. K. M., Ghosh, B., Nandagopal, K., and Jha, T. B. (2015). Morphological and Genetic characterization of Micropropagated Field grown plants of *Aloe vera* L. *Plant Tissue Culture and Biotechnology, 25,* 231-246.

Desjardins, Y., Tiessen, H., and Harney, P. M. (1987). The effect of sucrose and ancymidol on the *in vitro* rooting of nodal sections of asparagus. *Horticultural Science, 22,* 131-133.

Do, G. S., Seo, B. B., Ko, J. M., Lee, S. H., Pak, J. H., Kim, I. S., and Song, S-D. (1999). Analysis of somaclonal variation through tissue culture and chromosomal localization of rDNA sites by fluorescent *in situ* hybridization in wild *Allium tuberosum* and a regenerated variant. *Plant Cell, Tissue and Organ Culture, 57*, 113-119.

Dutta, Y. R., Gangopadhyay, G., Das, S., Dutta, B. K., and Mukherjee, K. K. (2003). Esterase as a marker to study the genetic fidelity of micropropagated banana. *Biologia Plantarum, 47,* 421-424.

Elsheikh, A. M., Daffalla, H. M. & Khalfala, M. M. (2013). *In vitro* micropropagation of the ornamental plant *Dieffenbachia* - A Review. *Universal Journal of Plant Science, 1,* 91-99.

Fereol, L., Chovelon, V., Causse, S., Michaux-Ferriere, N., and Kahane, R. (2002). Evidence of a somatic embryogenesis process for plant regeneration in garlic (*Allium sativum* L.). *Plant Cell Reports, 21,* 197-203.

Frabetti, M., Gutiérrez-Pesce, P., Mendoza-de Gyves, E., & Rugini, E. (2009). Micropropagation of *Teucrium fruticans* L., an ornamental and medicinal plant. *In vitro Cellular & Developmental Biology - Plant, 45,* 129-134.

Gantait, S., Mandal, N., Bhattacharyya, S., and Das, P. K. (2009). *In vitro* mass multiplication with genetic clonality in elephant garlic (*Allium ampeloprasum* L.). *Journal of Crop and Weed, 5,* 92-96.

Govil, S., and Gupta, S. C. (1997). Commercialization of plant tissue culture in India. *Plant Cell, Tissue and Organ Culture, 51,* 65-73.

Haisel, D., Hofman, P., Vágner, M., Lipavská, H., Tichá, I., Schäfer, C., and Čapková, V. (2001). *Ex vitro* phenotype stability is affected by *in vitro* cultivation. *Biologia Plantarum*, *44*, 321-324.

Haque, Sk. M., and Ghosh, B. (2013). High frequency microcloning of *Aloe vera* and their true-to-type conformity by molecular cytogenetic assessment of two years old field growing regenerated plants. *Botanical Studies, 54,* 46.

Jevremović, S., and Radojević, L. (2006). Establishment of efficient regeneration protocol from leaf explants of *Iris pumila* shoots cultured *in vitro*. *Scientia Horticulturae*, *108*, 100-103.

Keller, E. R. J., and Senula, A. (2013). Micropropagation and Cryopreservation of Garlic (*Allium sativum* L.). *In* M. Lambardi, E. A. Ozudogru, S. M. Jain (Eds.), Methods in Molecular Biology, Vol. 994: *Protocols for Micropropagation of Selected Economically-Important Horticultural Plants* (pp. 353-368). Humana Press.

Kozak, D., and Stelmaszczuk, M. (2013). Comparison of *Allium aflatunense* B. Fedtsch. 'Purple Sensation' and *Allium karataviense* Regel. 'Ivory Queen' regenerative capabilities in tissue culture. *Acta scientiarum Polonorum. Hortorum cultus, 12,* 197-213.

Larkin, P. J., and Scowcroft, W. R. (1981). Somaclonal variation - a novel source of variability from cell cultures for plant improvement. *Theoretical and Applied Genetics, 60,* 197-214.

Lee, S. H., Ryu, J. A., Do, G. S., Seo, B. B., Pak, J. H., Kim, I. S., and Song, S. D. (1998). Chromosome analysis by fluorescence in situ hybridization of callus-derived regenerants in *Allium cyaneum* R. *Plant Cell Reports*, *18*, 209-213.

Lee, K-S., and Ono, K. (1999). Chromosomal Variation in Callus Lines and regenerated Plantlets from Three Cultivars of *Allium fistulosum* L. (2n = 16). *Cytologia, 64,* 465-478.

Mante, S., and Tepper, H. B. (1983). Propagation of *Musa textilis* Neé plants from apical meristem slices *in vitro*. *Plant Cell, Tissue and Organ Culture, 2,* 151-159.

Mehrotra, S., Goel, M. K., Kukreja, A. K., and Mishra, B. N. (2007). Efficiency of liquid culture systems over conventional micropropagation: A progress towards commercialization. *African Journal of Biotechnology, 6,* 1484-1492.

Meyer, H. J., and van Staden, J. (1991). Rapid *in vitro* propagation of *Aloe barbadensis* Mill. *Plant Cell, Tissue and Organ Culture, 26,* 167-171.

Mukherjee, A., Sikdar, B., Ghosh, B., Banerjee, A., Ghosh. E., Bhattacharya, M., and Roy. S. C. (2013). RAPD and ISSR analysis of some economically important species, varieties, and cultivars of the genus *Allium* (Alliaceae). *Turkish Journal of Botany, 37*, 605-618.

Mukhopadhyay, K., and Roy, S. C. (1994). *In vitro* induction of 'runner' - a quick method of micropropagation in orchid. *Scientia Horticulturae, 56,* 331-337.

Murashige, T., and Skoog, F. (1962). A revised medium for rapid growth and bio assays with tobacco tissue cultures. *Physiologia Plantarum, 15*, 473-497.

Ottesen, C. (2014). Ornamental Alliums. *The American Gardener* September/October 2014: 12-17.

Pati, P. K., Sharma, M., Sood, A., and Ahuja P. S. (2005). Micropropagation of *Rosa damascena* and *R. bourboniana* in liquid cultures. In A. K. Hvoslef-Eide, and W. Preil (Eds.), *Liquid Culture Systems for in vitro Plant Propagation* (pp. 373-385). Springer: The Netherlands.

Peschke, V. M., and Phillips, R. L. (1992). Genetic implications of Somaclonal variation in plants. *Advances in Genetics, 30,* 41-75.

Poluboyarova, T. V., and Novikova, T. I. (2013). Clonal Micropropagation of Ornamental Onions of the Subgenus *Melanocrommyum* from Flower Organs. *Bull. State Nikitsky Bot. Gard, 107*, 46-51.

Ray, T., Saha, P., and Roy, S. C. (2005). *In vitro* plant regeneration from young capitulum explants of *Gerbera jamesonii. Plant Cell Biotechnology and Molecular Biology, 6,* 35-40.

Ray, T., Saha, P., and Roy, S. C. (2006a). Commercial production of *Cordyline terminalis* (L) Kunth. from shoot apex meristem and assessment for genetic stability of somaclones by isozyme markers. *Scientia Horticulturae, 108*, 289-294.

Ray, T., Dutta, I., Saha, P., Das, S., and Roy, S. C. (2006b). Genetic stability of three economically important micropropagated banana (*Musa* spp.) cultivars of lower Indo-Gangetic plains, as assessed by RAPD and ISSR markers. *Plant Cell, Tissue and Organ Culture, 85,* 11-21.

Ray, T., Saha, P., and Roy, S. C. (2013). Micropropagation of Cordyline terminalis. *In* M. Lambardi, E. A. Ozudogru and S. M. Jain, (Eds.), Methods in Molecular Biology, Vol. 994: *Protocols for Micropropagation of Selected Economically-Important Horticultural Plants* (pp. 269-277). Humana Press.

Razdan, M. K. (2003). *Introduction to Plant Tissue Culture, Second Edition*, Oxford & IBH Publishing Co., New Delhi, India.

Roy, S. C. (1980). Chromosomal variations in the callus tissues of *Allium tuberosum* and *A. cepa*. *Protoplasma, 102,* 171-176.

Roy, S. C., and Sarkar, A. (1991). *In vitro* regeneration and micropropagation of *Aloe vera* L. *Scientia Horticulturae, 47*, 107-113.

Ruffoni, B., Sacco, E., and Savona, M. (2013). *In vitro* Propagation of *Hydrangea* spp. *In* M. Lambardi, E. A. Ozudogru and S. M. Jain, (Eds.), Methods in Molecular Biology, Vol. 994: *Protocols for Micropropagation of Selected Economically-Important Horticultural Plants* (pp. 231-244). Humana Press.

Shen, X., Kane, M. E., and Chen, J. (2008). Effects of genotype, explant source, and plant growth regulators on indirect shoot organogenesis in *Dieffenbachia* cultivars. *In vitro Cellular & Developmental Biology-Plant, 44*, 282-288.

Sivanesan, I., Song, J. Y., Hwang, S. J., and Jeong, B. R. (2011). Micropropagation of *Cotoneaster wilsonii* Nakai-a rare endemic ornamental plant. *Plant Cell, Tissue and Organ Culture, 105,* 55-63.

Štajner, N. (2013). Micropropagation of *Asparagus* by *in vitro* shootculture. *In* M. Lambardi, E. A. Ozudogru, and S. M. Jain (Eds.), Methods in Molecular Biology, Vol. 994: *Protocols for*

Micropropagation of Selected Economically-Important Horticultural Plants (pp. 341-351). Humana Press.

Stelmaszczuk, M & Kozak, D. (2013). Micropropagation of *Allium neapolitanum* Cirillo. *Acta Scientiarum Polonorum, Hortorum Cultus, 12,* 193-206.

Šušek, A., Javornik, B., and Bohanec, B. (2002). Factors affecting direct organogenesis from flower explants of *Allium giganteum. Plant Cell, Tissue and Organ Culture,* 68, 27-33.

Vasil, I. K. (1985). Somatic embryogenesis and its consequences in the Gramineae. *In* R. R. Henke, K. W. Hughes, M. J. Constanin, and A. Hollaender (Eds.), *Tissue Culture in Forestry and Agriculture* (pp. 31-47). Plenum Press New York.

Waller, G. R., Mangiafico, S. & Ritchey, C. R. (1978). A chemical investigation of *Aloe barbadensis* Miller. *Proceedings of the Oklahoma Academy of Science, 58*, 69-76.

Wawrosch, C., Malla, P. R., and Kopp, B. (2001). Micropropagation of *Allium wallichii* Kunth, a threatened medicinal plant of Nepal. *In vitro Cellular & Developmental Biology- Plant, 37,* 555-557.

Yang, H. J., and Clore, W. J. (1973). Rapid vegetative propagation of asparagus through lateral bud culture. *Horticultural Science, 8,* 141-143.

Yang, H., Cheng J. C., and Kamada, H. (2000). Multiple pathways of Somatic embryogenesis at a High Frequency in *Asparagus officinalis* L. *Plant Biotechnology, 17,* 111-118.

In: Micropropagation of Ornamental Plants ISBN: 978-1-53614-541-0
Editor: Taoufik Bettaieb

Chapter 3

A New Wind Is Blowing Through Phalaenopsis Micropropagation

Stefaan P. O. Werbrouck*
Department of Plant and Crops, Faculty of Bioscience Engineering,
University Ghent, Ghent, Belgium

Abstract

Phalaenopsis hybrids are among the most important ornamental pot plants. Selected clones are mainly propagated by tissue culture. There exists a large genetic variation in the ability to grow, propagate and root *in vitro*. In this review a number of promising new plant tissue culture tools for *Phalaenopsis* will be discussed. Regarding shoot multiplication rate, unconventional topolin derived cytokinins, such as mFR, give higher micropropagation rates then BA, mT, MemT or MemTR, whereas during the last subculture, MemT is beneficial for rooting. Also the recipient plays a determining role. *Phalaenopsis* produces maximum shoots in tightly closed polypropylene recipients. Nevertheless, for an optimal quality, they should receive a final propagation step in a filter vessel that allows headspace ventilation. When monochromatic LED light was investigated, it turned out that red light stimulates the number of new shoots and roots, blue light gives dark green foliage and far red light

* Corresponding Author Email: Stefaan.werbrouck@ugent.be.

induces elongated light green leaves and reduces the number of roots. Also temporary immersion bioreactors offer interesting perspectives for cloning *Phalaenopsis*.

Keywords: orchid, cytokinin, bioreactor, light emitting diode

ABBREVIATIONS

B	blue LED;
BA	N_6 benzyladenine;
FR	far red LED;
MemT	*meta*-Methoxy topolin;
MemTR	*meta*-Methoxy topolin riboside;
mT	*meta*-topolin;
mTR	*meta*-topolin riboside;
mFR	*meta*-Fluoro-topolin riboside;
R	red LED

INTRODUCTION

Intensive *Phalaenopsis* breeding resulted in numerous valuable hybrids of extraordinary beauty which are highly appreciated as pot plants and cut flowers worldwide. Because these orchids have a monopodial growth habit, classical vegetative propagation is slow and difficult. Thanks to the micropropagation industry, nowadays *Phalaenopsis* takes the number one position of all micropropagated ornamentals. Among clones, there is a considerable genetic variation in their capacity to develop, propagate and root *in vitro*. Many valuable selections are rejected because of their poor or slow multiplication rate on mediums supplied with the traditional cytokinin BA. In this review new plant tissue culture tools are discussed that could speed up the multiplication rate without making concessions on shoot quality. They could potentially broaden the range of cultivars that can be commercialized.

***Meta*-Topolin Derived Cytokinins**

Important new tools came in with the development of a wide range of *meta*-topolin (mT) derivatives. *Meta*-topolin is a naturally occurring *meta*-hydroxylated benzyladenine with remarkable cytokinin effects (Werbrouck et al., 1996; Strnad et al., 1997; Aremu et al., 2012). To study their structure-activity relationship, Doležal et al., (2006, 2007) synthesized up to 48 mT derivatives with OH, Cl, F, OCH_3 substituents on the different positions of the aromatic ring. Their cytokinin activity was compared, with BA as standard compound, in three cytokinin bioassays: tobacco callus, wheat senescence and *Amaranthus* bioassay. Generally, the fluor containing derivatives were recognized as the most active compounds. The effects of a number of these topolins were examined on five recalcitrant *Phalaenopsis* genotypes. Partially defoliated rootless shoot explants were cultured on 5/8 strength MS-medium (Murashige and Skoog, 1962), 0.25 mg/lNAA, 20 g/l sucrose and 7 g/l agar (MC 29). This basal medium was supplemented with 10 µM of BA, mT, MemT, MemTR or mFR (Olchemim, Czech Republic). After 12 weeks all the new shoots were recorded. All the experiments were repeated three times with 30 explants per treatment (5 x 350 ml glass vessels with a nick in the rim x 6 plants per vessel). After the shoot multiplication phase, the shoots of one cultivar were transplanted on basal medium without cytokinins and the root formation was evaluated *in vitro*.

The results were presented in table 1. With exception of one genotype, all cultivars showed significantly more new shoots when treated with mFR (Figure 1). In this regard, MemT and its riboside MemTR were inferior to BA, mT and mFR. In none of the 5 cultivars, there was a significant difference between BA and mT regarding shoot number. Although there were more new shoots with mFR, these shoots were smaller than those formed on media containing the other cytokinins. Except for the size, there was no other visual difference in shoot quality between the treatments. Root formation occurred already in the multiplication phase, especially when MemT was used. Regarding shoot multiplication rate, these results suggest that mFR is an alternative for BA or mT. Plants cannot really cope with compounds with fluor. As halogenated compounds don't occur naturally in plants, it is hypothesized that mFR is probably broken down slower by cytokinin oxidases than other cytokinins (Aremu et al., 2012).

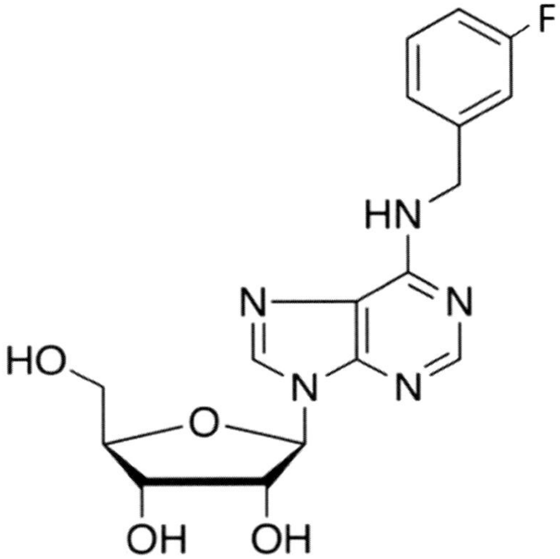

Figure 1. Chemical structure of mFR.

Table 1. Effect of 10 µM BA, mT, MemT, MemTR or mFR on multiplication coefficient of recalcitrant Phalaenopsis cultivars

Cytokinin	Cultivar Code				
	003	004	005	006	007
MemT	1.7 a	1.4 a	0.6 a	1.0 a	0.3 a
MemTR	2.2 a	1.7 ab	1.0 ab	1.9 ab	0.8 ab
BA	3.5 b	2.2 bc	1.2 b	2.0 ab	1.6 bc
mT	4.0 b	2.3 bc	1.3 b	2.5 b	2.0 c
mFR	5.1 c	3.9 d	2.1c	2.8 b	4.8 d

Means followed by the same letter are not significantly different (Tukey $p < 0,05$)

A major concern was whether a stable synthetic cytokinin such as mFR, or its metabolites, would inhibit root formation. The observations on cytokinin free medium indicated that there were no long lasting effects of the cytokinins. All plantlets formed roots, whatever the previously used cytokinin. The best rooting was performed by plants, which were previously developed on medium with MemT. Surprisingly, plants with a history of mT showed the poorest root formation. This is contradictory with the beneficial effects of mT on *Spathiphyllum* rooting (Werbrouck et al., 1996). The shoots with a BA history were remarkably heterogeneous. In conclusion, two compounds showed interesting effects in *Phalaenopsis*. First of all, mFR demonstrated the highest multiplication power. Secondly, MemT proved to be beneficial for root formation.

We suggest a sequence of media, where a number of shoot multiplication cultures on a medium with mFR (Figure 1), is followed by a subculture on medium with MemT. Chen (2016) calculated that by increasing the multiplication factor of *Phalaenopsis*from 2.0 to 2.5, the unit cost decreases with 26.2%. So, the effects of mFR are very promising to lower the production cost.

Initiation

Initiation can easily be accomplished starting from flower stalk or leave explants. Tanaka and Sakanishi (1978) showed that temperature should be risen to 28°C to obtain 100% vegetative buds, independent of their original position on the flower stalk. Adding 2.5 mg/lBA to the medium was found to be the optimal. Later, Ernst (1994) and Chen and Piluek (1995) reported that a higher induction efficiency was achieved with TDZ. Unlike flower stalk explants, foliar explants are easy to obtain and their availability is not restricted to any particular season. Protocorms can be formed from primordia developed on the surfaces of either small leaf fragments or whole immature leaves (Kuo et al., 2005). So, also on leaf explants, TDZ became a key growth regulator to induce protocorm-like bodies and adventitious shoots (Chen & Chang 2006). Interestingly, Pramanik et al., (2016) obtained a high number of secondary protocorm like bodies on half-strength MS when combining 0.5 mg/l BA and 2.5 mg/l mT. This allowed to avoid TDZ, with implies a decreased risk for somaclonal variation in *Phalaenopsis*.

Headspace Aeration

For many species, but especially for *Phalaenopsis*, the *in vitro* recipient needs special attention, as it plays a determining role in propagation rate and plant quality. Many different types of vessels are available, from rum bottles to sophisticated aerated models with filter lids that can even be manually adjusted. The question is if *Phalaenopsis* really needs headspace aeration in all phases. Next experiment is interesting in this regard (Werbrouck, unpublished results). A non-recalcitrant *Phalaenopsis* cv. was micropropagated on a 5/8 MS semi-medium, including 2% sucrose and the cytokinine mFR. The recipients were hermetically closed 565 ml polypropylene containers with an intact lid or with lids in which an 'L'= #10, 'XXL'= #30 or 'XXL+' = #40 filter (Microbox™, Belgium) was incorporated. This experiment was repeated three times and it was always observed that a maximal number of shoots were obtained in the tightly closed polypropylene recipients. The plants produced densely packed shoot clusters with a high number of shoots (table 2).

Figure 2. Effect of presence or absence of a filter in het lid of the recipient on the growth of Phalaenopsis *in vitro* from left to right: 'XXL+', 'XXL', 'L' and no filter (bar = 1 cm).

These shoots were generally small (Figure 2), but they could be separated easily. In the filter vessels, whatever the filter pore size, the number of shoots was smaller, but their size and plant quality was much better. This suggests that tightly closed containers are interesting for the propagation cycles of *Phalaenopsis*. But for an optimal plant quality, the shoots should receive a final propagation step in vessels that allow headspace aeration, such as Microbox™ 'L'. This would yield well sized shoots, ready for acclimatization.

Table 2. Effect of filter aeration level on the multiplication factor

Filter type	Multiplication factor
no filter	19.6 a
LARGE	6.8 b
XXL	9.3 b
XXL+	7.6 b

Means followed by the same letter are not significantly different (Tukey $p<0.05$)

Light Emitting Diodes

Energy-efficiency, ever reducing prices and a high photon flux have made LEDs a promising light source for micropropagation of *Phalaenopsis*. The currently available blue, red and far-red LED technology allows manipulating developmental parameters, such as elongation, rooting and leaf pigmentation. A major concern for commercial application of LED irradiation systems is the quality of *in vitro* plants grown under these light conditions. Indeed, when grown under (combinations of) monochromatic blue, red or far-red LED light remarkable effects can be observed in *Phalaenopsis*, compared to classical fluorescent light. This is demonstrated in following experiment.

Werbrouck (unpublished results) rooted a recalcitrant *Phalaenopsis* genotype on a 5/8 MS medium, 2% sucrose and 0,6% Plant Agar at 23±2°C in glass vessels without a filter but with a nick in the rim. The effects of blue (B, 454 nm), red (R 660 nm) and far red (FR 745 nm) light emitting diodes (40 μmol m^{-2} s^{-1} PAR) were examined. The culture racks were divided in boxes with white walls and a white door, in which Philips GreenPower LED Research Modules were mounted (Werbrouck et al.,

2011). Control treatment was cool fluorescent light (FL), provided by PHILIPS master TLD 36W 830 Reflex ECO (45 μmol m^{-2} s^{-1} PAR). Following observations were done: the highest multiplication coefficient was recorded under R, the highest level of anthocyanins was observed under B, the longest but also palest leaves under FR (Figure 3). Rooting was almost completely inhibited by FR, but it was stimulated by R.

Other reports about the effects of LED in *Phalaenopsis* are scarce. Wongnok et al., (2008) reported that axenic seed germination from five months old *Phalaenopsis*capsules yielded the best developed protocorms under 80% R + 20% B LEDs in Vacin and Went medium with 10 g/l sucrose. Chuang (2017) showed that after six weeks of culture, plants under R produced yellowish leaves with a low chlorophyll content but a high total sugar content and a large root length. Plantlets grown under R+B with a 5:2 R:B ratio showed a well activated carbohydrate metabolism and a better growth.

Temporary Immersion Bioreactors

Hempfling and Preil (2005) successfully used the twin bottle temporary immersion system of Escalona et al., (1999) for *invitro* shoot multiplication and rooting of *Phalaenopsis*. The highest shoot multiplication rate (25.4) was achieved after immersing every three hours with a modified half strength MS medium as published by Chen et al., (2000), supplemented with 2.2 μM TDZ.

The percentage of rooting and the number of roots was highest in medium with 5.7 μM IAA, applying an immersion every four hours. The fast growth and high multiplication rates in TIB are probably due to the uptake of nutrients and hormones over the whole plant surface. Nowadays, commercial *in vitro* companies experiment successfully with modern, more convenient temporary immersion bioreactor systems such as SETIS™ (Figure 3 and 4), but published data are currently missing, due to trade secrets.

Figure 3. Effect of red, blue and far red LEDs on leaf shape and color.

Figure 4. Proliferated Phalaenopsis in SETIS™, ready for harvest (courtesy of SETIS™).

Figure 5. Phalaenopsis shoot cluster produced in SETIS™ (courtesy of SETIS™).

Conclusion

In general, literature about modern *Phalaenopsis* micropropagation is scarce. It is big business and as a consequence, a lot of industrial protocols remain confidential. Nevertheless, a number of trends are evident: alternative cytokinins, tailor-made recipients for every *in vitro* phase, Light Emitting Diodes to shape plant morphology and Temporary Immersion Bioreactors to reduce manual labor. These new tools will shape the future production of this extraordinary ornamental that gives so much enjoyment to the consumer worldwide.

References

Aremu, A. O., Bairu, M. W., Doležal, K., Finnie, J. F., and Van Staden, J. (2012). Topolins: a panacea to plant tissue culture challenges? *Plant Cell, Tissue and Organ Culture (PCTOC), 108*(1), 1-16.

Chen, C. (2016). Cost analysis of plant micropropagation of *Phalaenopsis*. *Plant Cell, Tissue and Organ Culture (PCTOC), 126*(1), 167-175.

Chen, Y., and Piluek, C. (1995). Effects of thidiazuron and N6-benzyl-aminopurine on shoot regeneration of *Phalaenopsis*. *Plant Growth Regulation, 16*(1), 99-101.

Chen, Y. C., Chang, C., and Chang, W. C. (2000). A reliable protocol for plant regeneration from callus culture of *Phalaenopsis*. *In vitro Cellular & Developmental Biology-Plant*, *36*(5), 420-423.

Chuang, H. W. (2017). Light-Emitting Diodes Affect the Growth and Gene Expression of *Phalaenopsis* Plantlets *In Vitro*. *In Orchid Biotechnology III* (pp. 123-135).

Doležal, K., Popa, I., Hauserova, E, Spíchal, L., Chakrabarty, K., Novák, A., Kryštof, V., Voller, J., Holub, J., and Strnad, M. (2007). Preparation, biological activity and endogenous occurrence of N6-benzyladenosines. *Bioorganic & Medicinal Chemistry*, *15*, 3737–3747.

Doležal, K., Popa, I., Kryštof, V., Spíchal, L., Fojtíková, M., Holub, J., Lenobel, R., Schmülling, T., and Strnad, M. (2006). Preparation and biological activity of 6-benzylaminopurine derivatives in plants and human cancer cells. *Bioorganic & Medicinal Chemistry*, *14*, 875–884.

Ernst, R. (1994). Effects of thidiazuron on *in vitro* propagation of *Phalaenopsis* and *Doritaenopsis* (Orchidaceae). *Plant Cell, Tissue and Organ Culture, 39*(3), 273-275.

Escalona, M., Lorenzo, J. C., González, B., Daquinta, M., González, J. L., Desjardins, Y., and Borroto, C.G. (1999). Pineapple (*Ananas comosus* L. Merr) micropropagation in temporary immersion systems. *Plant Cell Reports, 18*(9), 743-748.

Hempfling, T., and Preil, W. (2005). Application of a temporary immersion system in mass propagation of *Phalaenopsis*. In A. K. Hvoslef-Eide and W. Preil (Eds.), *Liquid culture systems for in vitro plant propagation* (pp. 231-242). Springer: Dordrecht.

Kuo, H. L., Chen, J. T., and Chang, W. C. (2005). Efficient plant regeneration through direct somatic embryogenesis from leaf explants of Phalaenopsis 'Little Steve'. *In vitro Cellular and Developmental Biology-Plant, 41*(4), 453-456.

Murashige, T., and Skoog, F. (1962). A revised medium for rapid growth and bioassays with tobacco cultures. *Physiologia Plantarum, 15,* 473-497.

Pramanik, D., Yufdy, M.P., Shintiavira, H., and Winarto, B. (2016). *In vitro* Secondary Embryogenesis Derived from Meta-Topoline

Treatment on Mass Propagation of *Phalaenopsis* 'AMP 17'. *Notulae Scientia Biologicae, 8*(1), 62.

Strnad, M., Hanuš, J., Vaněk, T., Kamínek, M., Ballantine, J.A., Fussell, B., and Hanke, D. E. (1997). Meta-topolin, a highly active aromatic cytokinin from poplar leaves (*Populus* × *canadensis* Moench., cv. Robusta). *Phytochemistry, 45*(2), 213-218.

Tanaka, M., and Sakanishi, Y. (1978). Factors affecting the growth of *in vitro* cultured lateral buds from Phalaenopsis flower stalks. *Scientia Horticulturae*, *8*(2), 169-178.

Werbrouck, S., Buyle, H., Geelen, D., and Van Labeke, M. C. (2011). Effect of red-, far-red-and blue-light-emitting diodes on *in vitro* growth of *Ficus benjamina*. In *VII International Symposium on in vitro Culture and Horticultural Breeding, 961*, 533-538.

Werbrouck, S.P., Strnad, M., Van Onckelen, H.A., and Debergh, P.C. (1996). Meta-topolin, an alternative to benzyladenine in tissue culture? *Physiologia Plantarum*, *98*(2), 291-297.

Wongnok, A., Piluek, C., Techasilpitak, T., and Tantivivat, S. (2008). Effects of light emitting diodes on micropropagation of *Phalaenopsis* orchids. *In: International Workshop on Ornamental Plants 788*, 149-156.

In: Micropropagation of Ornamental Plants ISBN: 978-1-53614-541-0
Editor: Taoufik Bettaieb

Chapter 4

A NEW PROTOCOL FOR MICROPROPAGATION OF *LILIUM LONGIFLORUM* L.: USE OF CYCLODEXTRIN IN A DOUBLE PHASE MEDIUM

Sarra Arbaoui, Sihem Soufi, Soumaya Matmati and Taoufik Bettaieb*

Laboratory of Horticultural Sciences,
National Agronomic Institute of Tunisia,
University of Carthage, Tunisia

ABSTRACT

A new protocol has been developed in the present work for a large scale production *Lilium longiflurum* L. *In vitro* experiments showed that culture initiation is possible from scales on Murashing and Skoog (MS) (1962) medium without additional growth regulators. The best response,

* Corresponding Author Email: tbettaieb@yahoo.fr.

in terms of quantity, is obtained from explants of adult bulbs with 4.7 bulblets per scale. In the multiplication stage, 2 axillary shoots by bulblet were obtained in culture medium supplemented with 0.5 mg/l BAP and 0.2 mg/l IBA.

Addition of β-cyclodextrine, a protective substance to auxins, improved the *in vitro* rooting. Liquid medium supplemented with 6-benzylaminopuine (BAP) was added to form a double phase (solid/liquid) media. Indeed, the average number of roots increased from 5.64 to 7.20 roots per vitroplant in medium supplemented with β-cyclodextrin, and root length increased from 1.72 to 2.2 cm, compared to control.

Keywords: *Lilium,* micropopagation, bulbing, double-phase media, bulbscale

INTRODUCTION

Lilium is one of about 220 genus belonging to the Liliaceae family and comprises about 85 species, including economically relevant ornamental species (Feldmaier and McRae, 1982). It is one the most important cut flowers in the world because of its large attractive flowers, long vase life and aptitude to rehydrate after long transportation (Pandey et al. 2009).

Propagation of *Lilium* from seeds is not a practical method because of the very slow growth rate until flowering and the difficulty to obtain a large number of bulbs in a short time (Arzate-Fernández, 2002; Furyuya and Nomoura, 2004). Moreover, vegetative propagation from bulb scales has a very low rate, producing 3-4 bulblets per bulb scale and year, depending on the size and variety, which is not sufficient for large scale cultivation of this plant (Pandey et al. 2009). Thus, there is a need to develop a protocol for its mass propagation. Micropropagation is an advantageous alternative to conventional vegetative methods, since it not only increases several folds the multiplication rate, providing a continuous supply of bulblets, but also produces plants free from viruses and diseases. The successful use of *in vitro* culture for rapid propagation of *Lilium* has been reported, including *Lilium longiflorum* and oriental hybrid lilies (Lian et al. 2002; Bacchetta et al. 2003). The success of this technique depends on several factors like culture media, concentrations and combinations of plant growth regulators and explants source. Although many *Lilium* explants types have commonly been used, bulb scales have remained the best choice as explants to regenerate bulblets (Lian et al. 2003).

Several studies have focused on *in vitro* rooting that represents a major handicap to micropropagation methods for many species (Sekeli 2013; Ślesak et al. 2017). In the case of lily (*Lilium longiflorum* L.), *in vitro* rooting can occur successfully, although its enhancement is necessary for a better acclimatization of vitroplants or in the development of bulblets. Many investigations have shown that artificial growth hormones, especially auxins are constituted by molecules that are not very stable (George et al. 2008; Sauer et al. 2013). This can be overcome by using additives like β-cyclodextrin in *in vitro* culture. These compounds are cage molecules that include organic molecules like auxins forming a complex, and protecting them against photodegradation or oxidation. Complexes of cyclodextrin/auxin have no covalent bonds and can dissociate (Tsorteki et al. 2004).

This study aimed to evaluate the effect of plant growth regulators for the *in vitro* propagation of *Lilium longiflorum,* as well as the effect of cyclodextrin on its rooting and *in vitro* bulbing.

Culture Initiation and Shoot Multiplication

Explants used in the present study were taken from *Lilium longiflurum* L. Bulbs Amsterdam cultivar. Two sizes of different ages were considered: one-year old bulbs of 3 cm circumference and three-year old bulbs of 12 cm circumference. Bulbs were washed with distilled water and reduced into scales containing each a part of the basal plate. Scales were sterilized with 70% ethanol for few seconds followed by 10% sodium hypochlorite for 10 minutes and washed 5-6 times with sterilized distilled water before culturing.

The culture media used in *Lilium* micropagation was Murashige and Skoog (MS) (1962) supplemented with 6g/l agar and 30g/l sucrose. pH of all media was adjusted to 5.8 and then sterilized by autoclaving at 120°C under a pressure of 1 bar during 20 minutes.

For initiation and multiplication stages, the growth regulators indole-3-butyric acid (IBA) and benzylaminopurin (BAP) were added, either separately or in different combinations (Tables 1 and 2). Cultures were grown under a photoperiod of 16 hours, a luminous intensity of 36 µmole/m2/s provided by fluorescent tubes (Sylvania Gros Lux) and a temperature of 23°C ±1°C.

The effects of growth regulators and age of bulbs on culture initiation and bulblet regeneration are shown in Table 1. These results showed that culture initiation of *Lilium* depends on the age of the bulb used as well as the nature and concentration of the growth regulators added to the culture media. Explant response is expressed by bulblet formation at the scale base. The response of explants from mature bulbs was general, with exceptions due to some fungal contaminations. This experiment showed that the media promoting better explant response during this first micropropagation stage for the two sources of explants are media M1 containing 1 mg/l BAP and 0.2 mg/l IBA and M2 containing 2 mg/l BAP and 0.2 mg/l of IBA with a rate of 80%. However, explants from 1 year-old bulbs growing in control medium M0 had the highest response rate of 90% followed by explants growing in M1 containing 1 mg/l BAP and 0.2 mg/l IBA with a rate of 80%. Overall, these results show a significant capacity of *Lilium* for explant response and establishment in *in vitro* culture regardless of the growth regulators added to the culture media. These results were reported by Novak and Petru (1981) proving that the response of lily explants at this stage does not require growth regulators.

Following the explant recovery and after 4 weeks of culture, bulblets appear at scale bases (Figure 1). Explants from 3 year-old bulbs growing on medium M1 containing 1 mg/l BAP and 0.2 mg/l IBA regenerated 4.7 of bulblets per scale. Explants form 1 year-old bulbs grown in the same medium provided only 1.1 bulblets per scale. For the same plant species Novak and Petru (1981) showed the positive effect of MS Medium (1962) supplemented with 5 mg/l BAP associated with 1 mg/l IBA on *in vitro* bulblet neoformation. Moreover, Takayama (1990) reported that bulblet multiplication rate increased with the age of the explants.

After 6 weeks of initiation culture, the percentage of explant survival and the number of regenerated bulbs were recorded. For the multiplication stage, the following data was recorded after 8 weeks: multiplication rate, average number of axillary shoots generated per vitroshoot and average length of axillary shoots.

Table 1. Effects of plant growth regulators added to MS media and explants age on establishment of *in vitro* culture and bulb formation

Culture media	Growth regulators (mg/l)		Explant age (year)	Explant response (%)	Average number of regenerated bulblets
	BAP	IBA			
M0	0	0	1	90	0.4 ± 0.2
			3	30	1.3 ± 0.2[b]
M1	1	0.2	1	80	4.7 ± 0.8[a]
			3	80	1.1 ± 0.3[c]
M2	2	0.2	1	70	2.3 ± 1.1[b]
			3	80	0.5 ± 0.2[c]
M3	3	0.2	1	60	1.8 ± 0.9[b]
			3	40	0.1 ± 0.1[d]

Values followed by different letters are significantly different (Duncan test, 0.01).

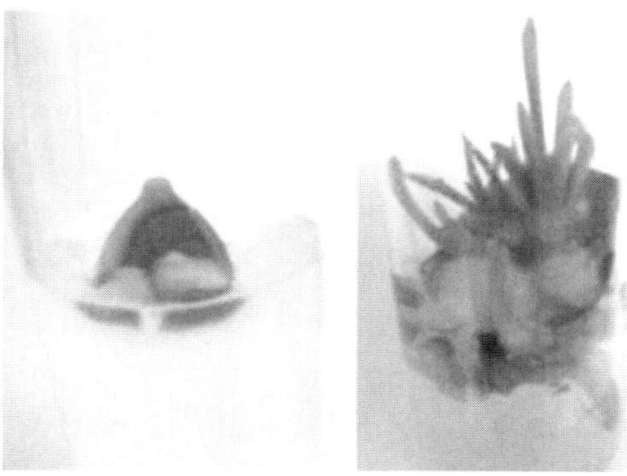

Figure 1. Bulblets regenerated at the base of *Lilium* scales after 6 weeks of culture.

Table 2. Effects of plant growth regulators added to MS media on axillary shoot formation

Culture medium	Growth regulators (mg/l)		Axillary shoot formation rate (%)	Number of auxillary shoots	Average shoot length (mm)
	BAP	IBA			
M0	0	0	58	0.9 ± 0.2^a	20.5 ± 1.2^b
M4	0.5	0.2	87.5	1.9 ± 0.4^b	18.7 ± 1.3^b
M5	1	0.2	60	1.4 ± 0.2^a	9.40 ± 1.2^a
M6	2	0.2	59	1.2 ± 0.1^a	8.20 ± 0.9^a
M7	3	0.2	60	1.5 ± 0.1^a	7.30 ± 1.1^a

Values followed by different letters are significantly different (Duncan test, 0.01).

Results of multiplication by axillary budding from vitrobulbs were significantly different depending on the combination of growth regulators added to the culture media (Table 2). 87.5% of vitroshoots grown in M4 medium, supplemented with 0.5 mg/l BAP and 0.2 mg/l IBA, regenerated auxiliary shoots.

For the culture media supplemented with 1, 2 and 3 mg/l BAP, axillary shoot formation was similar to that of the control media with a rate ranging between 50 and 60%, in spite of the well known positive effect of cytokinins on auxiliary shoot formation. The number of regenerated axillary shoots was significantly different among the different treatments during this stage of micropropagation (Table 2). M4 medium supplemented with 0.5 mg/l BAP and 0.2 mg/l IBA promotes this form of organogenesis with an average of 2 shoots by vitrobulb. The control and high cytokinin containing media have similar effects, with limited axillary budding. Therefore, it appears that the genus *Lilium* has a high multiplication potential and any excess in growth regulators reduces axillary budding.

The neoformed axillary shoots reach 20 mm after 6 weeks of culture in control medium M0. This can be explained by the effect of already existing endogenous hormones in the explants. The medium M1 containing 0.5 mg/l BAP and 0.2 mg/l IBA had a similar effect on the growth of vitroshoots and promoted the elongation with an average of 18.7 mm. This elongation is inversely proportional to the dose of cytokinin. Thus, an average length of 7 mm was recorded on the M7 medium that contained the highest dose of cytokinin.

For the rooting experiment, the vitroshoots used were regenerated by scale culture. The shoots produced are multiplied during five subcultures of four weeks each before transfer to rooting media.

IN VITRO ROOTING AND BULBING

For rooting and bulbing stages, the culture medium has a double-phase, solid and liquid (Figure 2). The solid phase of culture media used for rooting vitroplants, was supplemented with 30 g/l sucrose, solidified with 6g/l of agar and poured into 500 ml jars. The concentrations of IBA used were as follows: M0 = 0 mg/l; MR1 = 0.5 mg/l; MR2 = 1 mg/l; MR3 = 1.5 mg/l; MR4 = 2 mg/l.

The liquid phase used for bulbing and added to the solid media was supplemented with BAP and sucrose at different concentrations (Table 3) and poured, 30 days later, on the solid medium M4 that provides the best rooting results for the solid media. In order to investigate the impact of of β-cyclodextrin on *in vitro* rooting of *Lilium*, a second experiment was carried on, using the medium that provided the best results for rooting. Different doses of β-cyclodextrin were added to the rooting medium M4.

Figure 2. *Lilium* vitroshoots in double-phase medium (solid/liquid) for rooting and bulbing.

Table 3. Effects of IBA added to MS media on *in vitro* rooting of vitroshoots

Culture Medium	IBA(mg/l)	Rooting rate	Number of roots per vitroshoots	Average root length (cm)
M0	0	85	2.33 ± 0.19^{c}	0.75 ± 0.07^{d}
RM1	0.5	100	3.51 ± 0.24^{b}	0.92 ± 0.07^{c}
RM2	1	100	5.43 ± 0.20^{a}	1.22 ± 0.13^{b}
RM3	1.5	100	5.64 ± 0.27^{a}	1.72 ± 0.12^{a}
RM4	2	100	5.42 ± 0.19^{a}	1.63 ± 0.08^{a}

Values followed by different letters are significantly different (Duncan test, 0.01).

After 4 weeks of culture, the liquid medium M8 was added to M4, root development was evaluated after 4 weeks of culture in solid media only.

Root development was evaluated after 4 weeks of culture and bulb development was evaluated after 8 weeks of culture on the double-phase media, scoring rooting rate, number of roots per vitroshoot, average root length, bulbing rate and size of regenerated bulbs evaluated by bulb weight and circumferences.

Regardless of IBA concentrations in culture media, 100% rooting was obtained, while 85% rooting occurred in auxin-free media (Table 3). Vitroplant rooting in control medium is probably explained by a high content of endogenous auxins and exogenous application of IBA globally improves rhizogenesis. The number of neoformed roots on vitroplants depends on the culture medium. Indeed, the media RM2, RM3 and RM4 supplemented by 1; 1.5 and 2 mg/l of IBA, respectively, resulted in the development of about 5.5 roots per vitroplant. In the control media, the number of roots per vitroplant is 2.3. The longest root, exceeding 1.5 cm, was obtained on medium RM3 and RM4, containing 1.5 and 2 mg/l of IBA. Roots were shorter on the control medium, hormone free media, as well as on the RM1 and RM2.

Results of rooting experiments showed that addition of auxins to the culture media improved rooting of *Lilium*. Previous studies reported similar results related to rooting with IBA in *Lilium longiflorum* L. (Azadi and Khosh-Khui, 2007 and Matmati et al. 2014). They reported that *in vitro* rooting of lilies is possible in absence of auxins, and its addition contributed significantly to the enhancement of rooting rate, number and root length. Formation of lily bulbs was observed on all vitroshoots, but significant differences in weight and size of bulbs were recorded (Table 5). The highest weight and circumference of bulbs were obtained on the culture medium MB8 supplemented with 1 mg/l BAP and 45 g/l sucrose. A similar effect of *in vitro* bulbing of lily has been reported by several authors including those of Datu and Bhojwani (1995), Matmati et al. (2014). BAP-free media allows bulb formation but it does not appear to stimulate their growth. Moreover, addition of different sucrose concentrations to the culture media affects significantly *in vitro* bulbing of lily. The highest fresh weight and circumference were obtained with the higher concentrations of sucrose (4.5 and 6%) regardless of BAP concentration (Table 4).

However, sucrose concentration of 3% does not constitute a limiting factor for bulb formation. Although, the formed bulbs remain relatively small in size compared to those obtained on media supplemented with higher concentrations of sucrose. Nishiuchi (1980) reported that *in vitro* bulbing of tulip is improved by the addition of 4 to 6% of sucrose in culture media. Likewise, for lily, number and size of bulbs formed *in vitro* are highest when sucrose is used in culture media at a concentration of 9% (Takayama and Misawa, 1979 quoted by Datu and Bhojwani, 1995). For potatoes, *in vitro* tuberization was favored by 6% sucrose content (Espinosa et al. 1986)

According to the results presented in table 5, the presence of β-cyclodextrin in culture medium improved vitroshoot rooting. Moreover, number of roots increases from 5.64 to 7.20 roots per vitroplant, and root length increases from 1.72 to 2.2 cm. However, rooting is general even in the absence of β-cyclodextrin. The increase in root number and length had positive repercussions on bulb size; the mean weight increased in the presence of this additive from 0.31 g to 0.42 g, and the circumference from 2.10 to 2.60 cm.

Table 4. Effects of IBA and sucrose on *in vitro* bulbing after 60 days of culture on a double phase medium

Culture media	BAP (mg/l)	Sucrose (g/l)	Bulbing rate (%)	Bulb fresh weight (g)	Bulb circumference (cm)
MB1	0	30	100	0.17 ± 0.08^{e}	1.19 ± 0.05^{f}
MB2	0	45	100	0.18± 0.07^{e}	1.20 ± 0.09^{f}
MB3	0	60	100	0.17 ± 0.10^{e}	1.33 ± 0.04^{e}
MB4	0.5	30	100	0.28 ± 0.08^{b}	1.32 ± 0.08^{d}
MB5	0.5	45	100	0.26 ± 0.08^{c}	1.41 ± 0.12^{c}
MB6	0.5	60	100	0.24 ± 0.09^{d}	1.44 ± 0.14^{c}
MB7	1	30	100	0.27 ± 0.11^{b}	1.71 ± 0.07^{b}
MB8	1	45	100	**0.31 ± 0.05^{a}**	**2.10 ± 0.18^{a}**
MB9	1	60	100	0.28 ± 0.07^{b}	1.85 ± 0.10^{b}

Values followed by different letters are significantly different (Duncan test, 0.01).

Table 5. Effects of β-cyclodextrin on *in vitro* rooting and bulbing

Culture Medium	IBA (mg/l)	β-cyclodextrin (mg/l)	Number of roots per vitroshoot	Root length (cm)	Bulb fresh weight (g)	Bulb circumference (cm)
MC1	1.5	0	5.64 ± 0.27^{c}	1.72 ± 0.12^{b}	0.31± 0.05^{b}	2.10 ± 0.18^{b}
MC2	1.5	10	7.15 ± 0.12^{a}	2.18 ± 0.17^{a}	0.41 ±0.08^{a}	2.42 ± 0.15^{a}
MC3	1.5	20	7.20 ± 0.09^{a}	2.22 ± 0.10^{a}	0.39 ± 0.07^{a}	2.54 ± 0.09^{a}
MC4	1.5	30	7.01 ± 0.11^{b}	2.15 ± 0.15^{a}	0.42 ± 0.05^{a}	2.60 ± 0.11^{a}

Values followed by different letters are significantly different (Duncan test, 0.01).

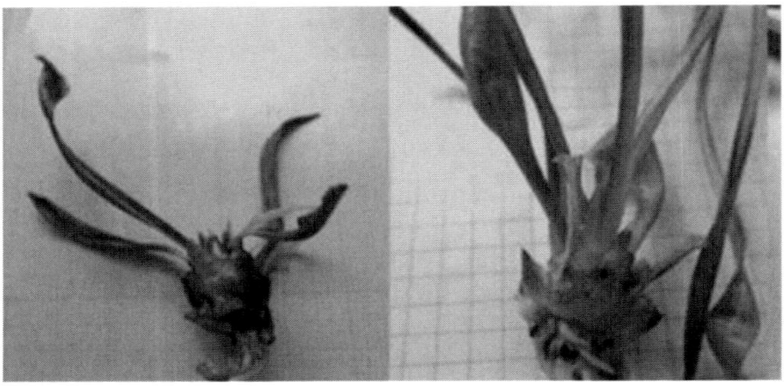

Figure 3. Vitroplants of lily (*Lilium longiflorum* L.) after bulbing.

According to the studies of Mura et al. (1995), presence of β-cyclodextrin in culture medium increases auxin solubility which positively affects *in vitro* rooting of olive vitroplants. Improvement effect of β-cyclodextrin for *in vitro* rooting has been also reported on jojoba (Apostolo et al., 2001), artichoke (Brutti et al. 2000; Dridi, 2003) and in *Nolina recurvata* Hemsl (Bettaieb et al. 2008).

Bellanger et al. (1992) explained that auxin is an unstable molecule that is protected by the β-cyclodextrin against photodegradation or oxidation and allows its slow and controlled release. Cyclodextrins are molecular cages able to include certain molecules and solubilize them through? Their cavity, thus forming complexes named inclusion complexes (Van der veen et al. 2000; Apostolo et al. 2001). Del Valle (2004) reported that the main force causing the formation of complexes, is the energy stabilization of the system by the replacement in the cavity of water molecules with high enthalpy of hydrophobic molecules which create apolar-apolar associations.

CONCLUSION

This study has resulted in a considerably improved micropropagation approach for *Lilium longiflurum* L with a great potentiality towards commercial propagation. *In vitro* multiplication experiments showed that 80-90% of scales form bulblets at their base on MS medium (1962) supplemented with 1mg/l BAP and 0.2 mg/l IBA. Each shoot obtained from microbulbs grown on MS medium enriched with 0.5 mg/l BAP and 0.2 mg/l IBA regenerates 2 axillary shoots.

On a double-phase culture medium (solid/liquid) for rooting and bulbing, vitroshoots of lily develop an average of 5.64 roots with 1.72 cm of length, and vitrobulbs of 0.31g and 2.10 cm of circumference. The double-phase culture medium consisted of mineral and vitamin additives of MS (1962), 1.5 mg/l IBA, 30 g/l sucrose and 6 g/l agar in its solid phase, and of a similar medium without agar for the liquid phase, in which the IBA was substituted for 1 mg/l BA and 45 g/l of sucrose were added. The addition of an auxin protective oligosaccharide, β-cyclodextrin, to the rooting medium at a concentration of 30 mg/l improves rooting and bulb growth. The vitroplants obtained have larger roots than those obtained from control medium, composed of 7.20 roots, with a length of 2.22 cm, and bulbs of 0.39 g of weight and a circumference of 2.54 cm. This protocol can be effectively used for *in vitro* multiplication of plants and bulbs of *Lilium longiflurum* L at commercial scale.

References

Apóstolo, N. M., Brutti, C., and Ferrarotti, S. A. (2001). Stimulation of root development with cyclodextrins on jojoba shoots *in vitro*. *In vitro Cellular & Developmental Biology - Plant,* 37, 414-418.

Arzate-Fernández, A. M., (2002). *In vitro* Culture for Propagation and Breeding of Lily (*Lilium* spp.). Universidad Autónoma del Estado de México. Colección: *Ciencias Naturales y Exactas*, serie: *Ciencias Agrícolas*. *Cigome. Toluca*, México. 96 p. [Autonomous Mexico State University. Collection: *Natural and Exact Sciences*, series: *Agricultural Sciences*. Cigome Toluca]

Azadi, P., and Khosh- Khui, M. (2007). Micropropagation of *Lilium ledebourii* Boiss as affected by plant growth regulator, sucrose concentration, harvesting season and cold treatments. *Electronical Journal of Biotechnology, 10,* 4-15.

Bacchetta, L., Remotti, P., Bernadini, C., and Saccardo F. (2003). Adventitious shoot regeneration from leaf explants and stem nodes of *Lilium*. *Plant Cell, Tissue and Organ Culture*, *74*(1), 37-44.

Bellanger, N., Djedaini-pilard, F., Berthault, P., and Perly, B. (1992). Cyclodextrines. *Phases Magazine* (9).

Bettaieb T., Mhamdi, M., and Hajlaoui, I. (2008). Micropropagation of *Nolina recurvata* Hemsl: β-cyclodextrin effects on rooting. *Scientia Horticulturae, 117,* 366–368.

Brutti, C., Apostolo, N. M., Ferrarotti, S. A., Liorente, B. E., and Krymkiewicz, N. (2000). Micropropagation of *Cynara scolymus* L.

emlpoying cyclodextrins to promote rhizodenesis. *Scientia Horticulturae, 8,* 1-10.

Dantu, K., and Bhojwani, S. (1995). *In vitro* corn formation and field evaluation of corm-derived plants of Gladiolus. *Scientia Horticulturae, 61*(1), 115-129.

Del Valle, E. M. (2004). Cyclodextrins and their uses: a review. *Process Biochemistry, 39,* 33–46.

Dridi, B. (2003). *An integrated micropropagation system for artichoke (Cynarascolymus L.).* PhD thesis in applied agronomic and biological sciences: cellular biotechnology and genetics. Belgium, 217 p.

Emsweller, L., and Uhring, J. (1964). *Interaction of temperature and growth regulator in overcoming self-incompatibility in Lilium longiflorum thumb.* Crops research division, Beltsville, Maryland, USA.

Espinoza, N. O., Estrada, R., Silva-Rodriguez, D., Tovar, P., and Lizarraga, R. (1986). The potato: a model crop plant for tissue culture. *Outlook Agriculture, 15,* 21-26.

Feldmaier, C., McRae J. (1982). *Lilien.* Ulmer: Stuttgart.

Furyuya, T., and Nomura, K. (2004). Rapid Production of *Lilium auratum* Bulbs from Zygotic Embryos. *Asia pacific journal of molecular biology and biotechnology, 12,* 39-42.

George, E. F., Hall, M. A., and De Klerk, G. J. (2008). Plant growth regulators I: Introduction, Auxins, their Analogues and Inhibitors. *In* Edwin, G. F., M. A. Hall, De Klerk, GJ. (Eds.), *Plant Propagation by Tissue Culture* (pp. 175-204). Springer: Dordrecht, The Netherlands.

Lian, M., Chakarabarty, D., and Paek, K. Y. (2003). Bulblet formation from bulb scale segments of *Lilium* using bioreactor system. *Biologia Plantarum, 46,* 199.

Lian, M. L., Murthy, H. H., and Paek, K. Y. (2002). Effects of light emitting diodes (LEDs) on the *in vitro* induction and growth of bulblets of *Lilium oriental* hybrid 'Pesaro.' *Scientia Horticulturae, 94*(3), 365-370.

Matmati, S., Soufi, S., Arbaoui. S., Bettaieb, T., and Mhamdi, M. (2013). Micropropagation of a bulbous ornamental: The lily (*Lilium longiflorum* L.). *Review of National Agronomic Institute of Tunisia, 28*(2), 15-25.

Mura, P., Ceccarelli, L., Faucci, M. T., Rinaldelli, E., and Mancuso, S. (1995). Improvement of solubility of indolbutyric acid by complexation with alphacyclodextrin and rhizogenic activity in *Olea europea* L. cv. Leccio del Corno. *Advance Horticultural Science,* 119-121.

Murashige, T., and Skoog, F. (1962). A revised medium for rapid growth and bioassays with tobacco tissue cultures. *Physiologia Plantarum, 15,* 473-497.

Nishiuchi, Y. (1980). Studies on vegetative propagation of tulips. IV. Regeneration of bulblets in bulb scale segments cultured *in vitro*. *Journal of the Japanese Society for Horticultural Science,* 235–40.

Novak, F., Petru, E. (1981). Tissue culture propagation of *Lilium* hybrids, *Scientia Horticulturae*, *14*, 191-199.

Pandey, R. K., Singh, A. K., and Mamta Sharma. (2009). *In vitro* Propagation of *Lilium*. *Biological Forum — An International Journal, 1*(2), 26-28.

Sauer, M., Robert, S., Kleine-Vehn, J. (2013). Auxin: simply complicated. *Journal of Experimental Botany*, *64*(9), 2565–2577.

Sekeli, R., Abdullah, J., Namasivayam, P., Muda, P., and Abu Bakar, U. (2013). Better Rooting Procedure to Enhance Survival Rate of Field Grown Malaysian Eksotika Papaya Transformed with 1-Amino-cyclopropane-1-Carboxylic Acid Oxidase Gene. *International Scholarly Research Notices: Biotechnology,* 10.

Ślesak, H., Liszniańska, M., and Ślesak, I. (2017). Rooting affects the photosystem II activity: *in vitro* and *ex vitro* studies on energy hybrid sorrel. *Acta Physiologia Plantarum, 39,* 210.

Takayama, S., and Ohkawa, K. (1990). Growth and flowering of *in vitro* propagated *Lilium auratum* bulbs in soil. *Plant tissue culture letters*, *3,* 187-192.

Tsorteki, F., Bethanis, K., and Mentzafos, D. (2004). Structure of the inclusion complexes of heptakis (2,3,6-tri-O-methyl)- β-cyclodextrin with indole-3-butyric acid and 2,4 dichlorophenoxyacetic acid. *Carbohydrate Research,* 339, 233–240.

Van der veen, B. A., Uitdehaag, J. C. M., Dijkstra, B. W., and Dijkhuize, L. (2000). Engineering of cyclodextrine glycosyltransferase reaction and product speci¢city. *Biochimica et Biophysica Acta, 1543*, 336-360.

In: Micropropagation of Ornamental Plants ISBN: 978-1-53614-541-0
Editor: Taoufik Bettaieb

Chapter 5

EFFICIENT PROTOCOLS FOR THE *IN VITRO* PROPAGATION OF THE BULBOUS ORNAMENTAL PLANT *GLADIOLUS GRANDIFLORUS* HORT

Sihem Soufi, Sarra Arbaoui and Taoufik Bettaieb*
Laboratory of Horticultural Sciences, National Agronomic Institute of Tunisia, University of Carthage, Tunisia

ABSTRACT

Gladiolus species are among of the most economically important ornamental plants cultivated all over the globe. They are widely demanded and own a horticultural value. Production of a large scale numbers of these valuable spices is possible by means of *in vitro* culture system. In the present investigation protocols were developed for *in vitro* propagation and large scale cormels production of two cultivars of *Gladiolus grandiflorus* Hort. Apical buds were used as explants. MS medium supplemeted with 0.5 mg l^{-1} IBA and 2 mg l^{-1} BA is ideal for shoot intiation and multiplication. For *in vitro* rooting and cormel

* Corresponding Author Email: tbettaieb@yahoo.fr.

formation an efficient method, cost and time effective method was developed using a double culture system with a solid phase with rooting media (MS supplemented with 0.5 mg l^{-1} of IBA media) and a liquid phase (MS Agar free medium supplemented with 6% of sucrose) poured 30 days on the top of the solid phase. The adopted culture system allowed concurrently within 60 days of culture the development of shoots, root cormels formation and had also a positive effect on cormels weight and circumferences of both *Gladiolus* cultivars.

Keywords: *Gladiolus grandiflorus* Hort, Peter pears, White friendship, cormels formation, *in vitro* propagation

INTRODUCTION

Gladiolus grandiflorus is a member of Iridaceae family and Ixidaceae subfamily and commonly called the "Queen" of bulbous flower, Sword lily and glad (Ahmad et al., 2007). It is a prominent bulbous cut flower prized and is considered as one among the commonly famous cut flowers cultivated and valued for its beautiful spikes in home gardening and landscape (Panday et al., 2010; Luo et al,. 2012). Singh et al., (2017) reported that being an important bulbous ornamental plant in cut flowers industry, *Gladiolus* actually occupied the fifth position in the world cut flower trade.

Conventionally, *Gladiolus* is mainly propagated either by corms or cormels (Nagarju and Parthasarathy, 1997). However, this propagation technique is lengthy for large-scale multiplication and limited due to several factors including the non-availability of a large quantity of corms for commercial use, the low multiplication and growth rate of corms, the physiological dormancy of the corm and cormels and corm rot during storage (Singh and Dohare, 1994; Priyakumari and Sheela, 2005). To overcome those limiting factors, an efficient multiplication method is required using *in vitro* propagation. The present studies were undertaken in order to develop a cost and time-effective *in vitro* propagation scheme for cormel production of *Gladiolus* cultivars.

ESTABLISHMENT OF CULTURES AND SHOOTS INDUCTION

The experiments were carried out on two cultivars of *Gladiolus grandiflorus* Hort. (Peter pears and White friendship). Apical bud explants obtained directly from surface sterilized corms were used in this experiment. Apical buds explants were trimmed down until the size ranged from 2.0 to 3.0 mm and then inoculated on a Murashige and Skoog (1962) (MS) basal medium for culture initiation containing 30.0 g l^{-1} sucrose and 6 g l^{-1} agar. Media pH were adjusted to 5.8 prior to autoclave at 121°C for 20 minutes. Different concentrations and combinations of IBA (Indole-3-butyric acid) and BA (benzylaaminopurine) were incorporated in each medium for induction of buds from explants:

- Medium 1 (hormone free)
- Medium 2 (1 mg l-1 IBA)
- Medium 3 (1 mg l-1 BA)
- Medium 4 (2 mg l-1 BA and 0.5 mg l-1 IBA)
- Medium 5 (2 mg l-1 IBA and 0.5 mg l-1 BA)

These cultures were maintained in a growth chamber with 16 h photoperiod, 25 ± 2 °C temperature under light (36 $\mu mol.m^{-2}.s^{-1}$) provided by fluorescent tubes. All cultures were examined periodically and the morphological changes were recorded after four weeks of culture.

The effect of different concentrations and combinations of BA and IBA on the development of apical buds of two *Gladiolus* cultivars 'Peter pears' and 'White friendship' was therefore investigated. The obtained results (Table 1) show a difference in explants response to culture media. During shoot induction period and after 4 weeks of apical buds inoculation on hormone-free MS and on MS supplemented with 1 mg l^{-1} of IBA, rooting of 'Peter pears' and 'White friendship' was completely achieved (100%) (Table 1). These findings are partially in agreement with the investigation of Haman et al., (1995). In the same medium callus induction was also observed. Ziv et Lilien-Kipnis (1990) reported that among several plant growth regulators tested, NAA and 2,4-D supplemented at high concentrations to MS medium led to the formation of callus. Kabir et al. (2014) have also tested different various plant growth regulators and have concluded that 7.5 mg l^{-1} of NAA is the suitable plant growth regulator and concentration for a better callus induction in *Gladiolus dalenii*. The moderate callus formation observed in our study is attributed to the endogenous auxin and IBA supplementation to the culture media. Among the various concentrations and combinations of BA and IBA tested, Medium 3 supplemented with 1 mg l^{-1} BA and Medium 4 supplemented with 0.5 mg l^{-1} IBA and 2 mg l^{-1} BA were used in the next multiplication media. Caulogenesis was pronounced for both cultivars and this is particularly due to the dominance of BA in MS media as explained by the relatively high buds percentage. On these media both cultivars exhibited the same organogenesis behavior. The present results highlight the predictable growth responses of *Gladiolus* shoots inoculated *in vitro*.

Shoots Multiplication

Based on the rate of axillary bud and the quality of produced shoots, MS Media supplemented respectively with 1 mg l^{-1} BA (Medium 3) and 2 mg l^{-1} BA and 0.5 mg l^{-1} IBA (Medium 4) were found to be optimal for cultures multiplication. After four weeks of growth under growth conditions as described above, growth parameters such as multiplication rate and leaves presence were measured. During the shoot multiplication stage, both *Gladiolus* cultivars show similar behavior during the establishment phase. Various concentration and combination of auxin and cytokinin tested did not significantly alter the rate of multiplication during the first subculture which lasted four weeks. About 5 shoots were obtained from a single explants cultured and this rate of multiplication did not present any variation during repetitious subcultures (Figure 1). It is worthy to note that shoots with a height superior to 0.5 cm were only observed on MS supplemented with 2 mg l^{-1} BA and 0.5 mg l^{-1} IBA. It seems that the presence of auxin in the medium was found to be necessary to promote shoots elongation.

Table 1. Effects of different concentrations and combinations of cytokinin and auxin on shoot multiplication of *Gladiolus gradiflorus* Hort. Cultivars from apical bud explants

BA (mg l^{-1})	IBA (mg l^{-1})	Gladiolus Cultivars	Roots percentage (%)	Callus percentage (%)	Buds percentage (%)
-	-	PP	100	16	10
-	-	WF	100	15	5
-	1	PP	100	25	8
-	1	WF	100	30	6
1	-	PP	0	0	100
1	-	WF	0	2	100
2	0.5	PP	0	0	100
2	0.5	WF	0	0	100
0.5	2	PP	45	14	11
0.5	2	WF	36	16	4

PP= Peter pears, WF= White friendship.

Table 2. Effects of different concentrations and combinations of cytokinin and auxin on shoot multiplication of *Gladiolus gradiflorus* Hort. Cultivars from apical bud explants

BA (mg l-1)	IBA mg l-1)	Cultivars	Multiplication rate (%)	Shoot length (%) (> 0.5 cm)	Shoot length (%) (< 0.5 cm)
1	-	PP	4.6	60	40
1	-	WF	4.4	50	50
2	0.5	PP	4.9	0	100
2	0.5	WF	4.8	0	100

PP= Peter pears, WF= White friendship.

In vitro Rooting and Cormels Formation

In vitro Rooting of Shoots

For the rooting of *in vitro* produced shoots, MS media mineral salt were diluted by half and each media were supplemented respectively with auxin (IBA) at different concentrations (0, 0.1, 0.5, 1 and 5 mg l-1), 2 mg l-1 of charcoal, 6 g of agar was used to solidify media. The effects of different concentration of auxin (IBA) on the root induction of Gladiolus cultivars were determined by measuring the root frequency, root length, and root number of each treatment after 4 weeks days of incubation.

Based on the obtained results presented in table 3, rhizogenesis was mostly undergone in all inoculated shoots and those independently of IBA concentration. These results demonstrate that the plantlets could produce roots in MS medium devoid of or with auxins. Our findings suggested that rooted plantlets on MS medium devoid with auxin could be attributed to the endogenous amount of auxin initiate root induction. However, the mean response of two Gladiolus cultivars to various levels of IBA concentration was shown to be highly significant ($P < 0.01$) for root number as well as for root length. Among the various concentrations of IBA tested, the highest number of roots was produced in MS medium supplemented with 0.5 and 1 mg l-1 of IBA and the longest root was achieved on MS medium containing 1 mg l-1 IBA. It is noteworthy that number of roots per shoots deceased when IBA concentration reached 5 mg l-1. It seems that high concentrations of IBA are harmful for *in vitro* rooting of Gladiolus cultivars.

In vitro Cormels Formation

To evaluate *Gladiolus* cormels formation after 60 days of inoculation a double phase culture system was used. This double phase system is composed two different phases (Solid and liquid) with two different media composition. The solid phase consisting of rooting media (details above) placed at the bottom of the vessel and above which a liquid phase composed of MS basal (MS media mineral salt were diluted by half), supplemented with 0.5 mg l^{-1} IBA and different concentration of sucrose (0; 1.5; 3; 6 and 8%) consisting of cormels formation media. After 30 days and under sterile condition the liquid phase was poured into the same vessel.

(a)

(b)

Figure 1. Effects of plant growth regulator on *Gladiolus grandiflorus* shoot multiplication. (a) Development of high quality plantlet from apical bud explants of on multiplication media (MS medium with 2 mg l-1 BA and 0.5 mg l-1 IBA) (b) microshoots from incised shoot bud explants.

Table 3. Effects of different concentrations of auxin on *in vitro* cormels formation of *Gladiolus gradiflorus* Hort. cultivars after 60 days of on inoculation on double phase system culture

IBA (mg l^{-1}) (mg/l)	Cult-ivars	Rooted shoots (%)	Roots num-ber	Height average of roots (cm)	Cormels formation (%)	Cormels fresh weight (g)	Cormels cicumferen-ces (cm)
0	PP	100	3.1^{d}	1.5^{ab}	100	0.24^{c}	2.1^{bc}
	WF	100	2.9^{d}	0.9^{c}	100	0.26^{bc}	2.3^{b}
0.1	PP	100	3.5^{c}	0.9^{c}	100	0.3^{ab}	2.4^{ab}
	WF	100	3.3^{cd}	1.1^{bc}	100	0.29^{ab}	2.2^{bc}
0.5	PP	100	5.4^{a}	1.2^{bc}	100	0.36^{a}	2.9^{a}
	WF	100	5.7^{a}	1.4^{b}	100	0.34^{a}	2.8^{a}
1	PP	100	5.6^{a}	1.7^{a}	100	0.34^{a}	3.1^{a}
	WF	100	5.4^{a}	1.5^{ab}	100	0.33^{a}	2.9^{a}
5	PP	100	3.6^{c}	1.3^{b}	100	0.31^{ab}	2.9^{a}
	WF	100	4.9^{b}	1.3^{b}	100	0.35^{a}	2.7^{ab}

Different letters indicate significant differences between media at $P = 0.01$. $n = 24$. Cv = Cultivars, PP = Peter pears, WF = White friendship.

The use of different concentrations of IBA had showed significant ($P < 0.01$) effects on corm fresh weight and circumference as shown in Table 3 after 60 days of culture. Corm formation occurred in all tested media including control media but it does not enhance their growth. In the present study, the best concentration of IBA for rooting and corm formation for both *Gladiolus* cultivars was 0.5 mg l^{-1} of IBA and which is used in the next corm formation assays. The addition of sucrose at different concentration had a positive and a significant effect on *Gladiolus* cultivars fresh weight and corm circumferences as shown in table 4. Corm formation occurred also in all tested media including control. High corm fresh weight and high circumferences for both cultivars were recorded on MS media supplemented with 6 and 8% of sucrose. MS media devoid or with low concentration of sucrose (1.5%) did not prevent *in vitro* corm formation but it constituted a limiting factor for their growth. Several studies reported that sucrose plays a crucial role in *Gladiolus in vitro* corm formation which confirms our findings. Other studies reported that a relatively high concentration of sucrose in the medium are needed to increase growth of tuberous organs (Mares et al., 1985; Nagaraju et al., 2002). The importance of relatively high concentration of sucrose for *in vitro* corm formation has been reported for several species. Nishiuchi Y., 1980 found that in presence of 4 to 6% of sucrose *in vitro* corm formation of tulip was promoted. Espinosa et al., (1986) and Bettaieb et al. (2002) found that *in vitro* tuberisation of potato was enhanced by the supplementation of 6% of sucrose Nagaraju et al., (2002) found that Supplementation of 10 mg l^{-1} paclobutrazol and 120 mg l^{-1} sucrose to MS medium favored formation of bigger cormels. Memon et al. (2009) obtained three different sizes of cormel from rooted shoots cultured in MS medium supplemented with higher levels of sucrose (7%) and lower levels of IBA (1 mg l^{-1}) in variety White Friendship.

Table 4. Effects of different sucrose concentrations on *in vitro* cormels formation of *Gladiolus gradiflorus* Hort. cultivars after 60 days of inoculation on double phase system culture

Sucrose (%)	Gladiolus Cultivars	Cormel formation (%)	Cormels fresh weight (g)	Cormels circumferences (cm)
0	PP	100	0.14^{d}	0.9^{d}
0	WF	100	0.16^{d}	1.0^{d}
1.5	PP	100	0.28^{c}	2.1^{c}
1.5	WF	100	0.30^{c}	2.0^{c}
3	PP	100	0.36^{b}	2.9^{b}
3	WF	100	0.34^{b}	2.8^{b}
6	PP	100	0.41^{a}	3.2^{a}
6	WF	100	0.39^{ab}	3.3^{a}
8	PP	100	0.42^{a}	3.1^{ab}
8	WF	100	0.40^{a}	3.2^{a}

Different letters indicate significant differences between media at $P = 0.01$. $n = 24$. PP= Peter pears, WF= White friendship.

CONCLUSION

The overall results demonstrated that both *Gladiolus* cultivars could successfully multiplied from apical buds obtained from corm. They also demonstrated that growing media composition plays a key role in the establishment of a successful *in vitro* micro propagation and cormel production. Our findings highlight that production of high quality and healthy vigorous shoots. Almost all the obtained shoots were achieved rooting and cormel formation using *in vitro* a double phase culture system. It is worthy to conclude that an efficient *in vitro* propagation and corm formation protocol were developed for *Gladiolus* cultivars in the present study. The developed protocol is time and cost effective in terms of production of multiple cormels in short time span.

Figure 2. Roots and cormels formation of *Gladiolus grandiflorus* cultivars on a double phase culture system after 60 days of inoculation on MS media. (a) Solid phase for rooting media composed of MS with 0.5 mg l-1 BA, (b) Liquide phase for cormels formation composed of MS with 0.5 mg l-1 BA and 6% sucrose.

REFERENCES

Bajaj, Y. P. S., Sidhu, M. M. S., and Gill, A. P. S. (1992). *Micro-propagation of Gladiolus*. Biotech-Agri-Berlin. Germ.: Springer-Verlag. 19, 35-143. 2.

Bettaieb, T., Hannachi, C., M'hamdi, M., and Ben Kheder, M. (2002). Improvement of *in vitro* tuberization of the potato (*Solanum tuberosum* L.) Var. Spunta. *Review of the National Agronomic Institute of Tunisia, 17*(1), 69-79.

Dantu, P. K., and Bhojwani, S. S. (1995). *In vitro* corm formation and field evaluation of corm-derived plants of *Gladiolus*. *Scientia Horticulturae, 61,* 115- 129.

Espinosa, N. O., Estrada, R., Silva-Rodriguez, D., Tovar, P., and Lizarraga, R. (1986). The potato: a model crop plant tissue culture. *Outlook on Agriculture, 15,* 21-26.

Hamann, H. (1995). Contribution to the development of a gene transfer technique via *Agrobacterium tumefasciens* in gladiolus (*Gladiolus grandiflorus* Hort.). *Master's degree in chemistry and plant biology.* Rennes University 1. France 20 p.

Hussey, G. (1977). *In vitro* propagation of *Gladiolus* by precocious axillary Références bibliographiques shoot formation. *Scientia Horticulturae, 6,* 287-296.

Lilien-Kipnis, H., and Kochba, M. (1987). Mass propagation of new *Gladiolus* hybrids. *Acta Horticulturae, 212*(8), 631-638.

Luo, X., Yi, J., Zhong, X. H., Lian, Q. L., Khan, M.A., Caoa, X., Li, X. X., Gao, M. W., Wu, J., Chen, J., and Yi, M. F. (2012). Cloning, characterization and expression analysis of key genes involved in ABA metabolism in *Gladiolus* cormels during storage. *Scientia Horticulturae, 143,* 115-121.

Mares, D. J., Sowokinos, J. R., and Hawker, J. S. (1985). Carbohydrate metabolism in developing potato tubers. In P. H. Li (Eds.). *Potato physiology* (pp. 279-327). Academic Press: Orlando.

Murashige, T., and Skoog, F. (1962). A revised medium for rapid growth and bioassays with tobacco cultures. *Physiologia plantarum, 15*, 473-497.

Nagaraju, V., Bhowmik, G., and Parthasarathy, V. A. (2002). Effect of paclobutrazol and sucrose on *in vitro* cormel formation in *gladiolus*. *Acta Botanica Croatica, 61,* 27-33.

Nishiuchi, Y. (1980). Studies on vegetative propagation of tulips. Regeneration of bulblets in bulb scale segments cultured *in vitro*. *Journal of the Japanese Society for Horticultural Science, 49,* 235-240.

Pandey, R. K., Sheetal, D., Pal, P. S., Shivani, J., and Bhat, D. J. (2010). Performance of *Gladiolus* cultivars under Jammu conditions. *Journal of Plant Research, 9*(2), 210-214.

Steinitz B., Cohen A., Goldberg Z., and Kochba M. (1991). Precocious corm formation in liquid shake cultures. *Plant Cell, Tissue and Organ Culture, 26,* 63- 70.

Ziv, M. (1979). Transplanting *Gladiolus* plants propagated *in vitro*. *Scientia Horticulturae, 11*, 257-260. 12.

Ziv, M., and Lilien-Kipnis, H. (1990). *Gladiolus*. In P. V. Ammirato, D. R. Evans, W. R. Sharp, and Y. P. S. Bajaj (Eds.), *Handbook of plant cell culture*, volume 5 ornamental species: McGraw-Hill Publishing Company: New York.

Part III: *In vitro* Propagation of Some Woody Ornamental Plants Via Organogenesis and Somatic Embryogenesis: Generalities and Studied Cases

In: Micropropagation of Ornamental Plants
Editor: Taoufik Bettaieb
ISBN: 978-1-53614-541-0

Chapter 6

MICROPROPAGATION OF JACARANDA (*JACARANDA MIMOSAEFOLIA* D. DON)

Tsuyoshi E. Maruyama*[*] *and Yoshihisa Hosoi
Forestry and Forest Products Research Institute (FFPRI),
Department of Forest Molecular Genetics and Biotechnology,
Tsukuba, Ibaraki, Japan

ABSTRACT

Jacaranda (*Jacaranda mimosaefolia* D. Don.) was regenerated by shoot-tip subcultures on medium supplemented with large concentrations of kinetin and solidified with Gelrite. For shoot elongation, the reduction of the cytokinin level and the addition of activated charcoal to the medium were optimal. High rooting frequencies were achieved on medium supplemented with indole-3-butyric acid alone, or in combination with naphthaleneacetic acid. Regenerated plantlets were transferred to vermiculite substratum and acclimatized successfully in a growth cabinet. Alginate-encapsulation and germplasm conservation approaches are also described.

[*] Corresponding Author Email: tsumaruy@ffpri.affrc.go.jp.

Keywords: bignoniaceae, tropical trees, *in vitro* propagation, shoot-tip culture, artificial seeds, germplasm conservation

INTRODUCTION

Jacaranda mimosaefolia D. Don. (Jacaranda), endemic to the Peru–Amazon region, is one of the most beautiful ornamental trees across the world and is planted on a large scale in gardens, parks, avenues, and in-house gardens in tropical and subtropical countries (Figure 1). Jacaranda is traditionally propagated through seeds in nurseries. Although sexual propagation is important for introducing new genes to prevent inbreeding and for achieving genetic gain for those characteristics controlled by additive gene effects (Thorpe et al. 1991), clonal propagation of high-quality trees is also desirable. Clonal propagation is especially important in ornamental trees where aesthetic appeal, blossom, and flower color and resistance to pests and diseases are important characteristics to be retained in the progeny (Bajaj 1989). Micropropagation is one of the true-to-type propagation methods of selected genotypes using *in vitro* culture (Debergh and Read 1991). The aim of the present work was to investigate the optimal conditions for the propagation of *J. mimosaefolia*.

Figure 1. *Jacaranda mimosaefolia* planted as an ornamental tree.

In this review, based on previous reports (Maruyama et al. 1993, Maruyama et al. 1997a, b, Maruyama 2003), we describe the conditions for micropropagation of Jacaranda by shoot-tip cultures and also for the alginate-encapsulation of shoot tips and germplasm conservation at temperature conditions above freezing.

MICROPROPAGATION

Culture Initiation

Seeds were surface-sterilized by agitating them in 70% (v/v) ethanol for 3 min and additionally in 5% (w/v available chlorine) sodium hypochlorite solution for 10 min. After three rinses with sterile distilled water, the seeds were placed on 0.8% (w/v) agar medium in 300-ml Erlenmeyer flasks. For initial culture, shoot tips of aseptically germinated seedlings were planted on B5 medium (Gamborg et al. 1968) supplemented with four different concentrations of 6-benzylaminopurine (BAP) (Table 1). Data were collected after 45 days of culture under a light intensity of 65 µmol m^{-2} s^{-1}. Thereafter, two subcultures (at 45-day intervals) were conducted on B5 medium containing 2 µM BAP and 2 µM trans-zeatin (ZEA).

Table 1. Effects of different concentrations of BAP on the initial culture of shoot tips excised from aseptically germinated seedlings of *J. mimosaefolia*

Concentration of BAP (µM)	Cultured explants	Explants that formed shoots
1	2	1
2	3	3
4	3	3
10	3	2

All the seeds were successfully surface-sterilized. Six weeks after sowing, the percentage of germination was 14% and the number of shoot tips obtained from germinated seedlings was 11 shoots. The 11 shoot tips obtained from 1- to 2-month-old aseptically germinated seedlings were cut into lengths of 1–2 cm and then cultured on B5 medium supplemented with BAP at four concentrations as shown in Table 1. Creamy-colored calli were formed at the bases of explants in contact with the medium. After two consecutive subcultures on B5 medium supplemented with 2 μM BAP and 2 μM ZEA, a total of 175 shoots were obtained from the 11 initial apical shoots (three- to four-fold propagation rate per subculture).

Effects of Basal Media

The effects of basal media on survival, shoot multiplication, and shoot growth were tested. Shoot-tip explants were cultured on MS medium (Murashige and Skoog 1962), WP medium (Lloyd and McCown 1981), BT medium (Chalupa 1981), and B5 medium containing 4.44 μM BAP. The pH was adjusted to 5.8 before autoclaving at 1.1 kg.cm^{-2} and 121°C for 20 min. The cultures were maintained at 25°C under a light intensity of 65 μmol m^{-2} s^{-1} provided by cool white fluorescent lamps (100 V, 40 W; Toshiba, Tokyo, Japan) for a 16-h photoperiod. Data were collected from 10 cultured shoots for each medium after 30 days of culture.

Table 2. Effects of different basal media on survival rate, shoot multiplication, and shoot growth from the shoot tips of *J. mimosaefolia*

Medium	Survival rate (%)	Shooting rate (%)	Shoot per explant Mean ± SE	Maximum length of shoots (cm) Mean ± SE
MS	100	100	2.7 ± 0.42	1.3 ± 0.17
WPM	100	100	2.6 ± 0.37	1.7 ± 0.28
BTM	100	100	2.0 ± 0.21	1.5 ± 0.20
B5	100	100	3.3 ± 0.50	2.2 ± 0.23

SE: standard error.

Table 2 shows the results of survival, shoot multiplication, and shoot growth after 30 days of culture. The survival rate was 100% in the four tested media. However, the best result with regard to shoot multiplication and growth of shoots was achieved in B5 medium. There were no significant differences among the other media.

Effects of Cytokinins

To determine the effects of different cytokinins on survival, shoot multiplication, and shoot growth, shoot tips from subcultured explants were cultured on B5 medium containing different cytokinins, namely, BAP, ZEA, and 6-furfurylaminopurine (KIN) at three concentrations (Table 3). Data were collected from 10 cultured shoots for each of the tested cytokinin and concentration after 40 days of culture on B5 medium under a light intensity of 65 µmol m^{-2} s^{-1} for a 16-h photoperiod.

New shoot tips obtained by subculture were cultured on B5 medium supplemented with three concentrations (1, 10, and 100 µM) of BAP, ZEA, or KIN. Table 3 shows the results of survival, shoot multiplication, and shoot growth after 40 days of culturing. Even though cytokinin concentrations are dependent on the tissue and chemical employed, in general, the concentration ranging from 0.5 to 10.0 mg l^{-1} is most suited for plant propagation (Holdgate 1977). However, Murashige indicates that a high concentration of KIN (30 mg l^{-1}) is required for *Gloxinia* (Murashige 1974). In this study, for multiple shoot induction by shoot-tip cultures, the best result (6.1 shoots per explant, in average) was obtained with the highest level of KIN (100 µM) (Figure 2). Addition of BAP or ZEA at the same concentration level resulted in browning or death of explants. When BAP or ZEA was supplemented at the 10 µM level, 3.4 and 2.8 shoots per explant in average were produced, respectively. In both cases, calli were formed first at the bases of the explants, and new shoots were formed through the calli. BAP levels higher than 10 µM (25 and 50 µM) and KIN levels higher than 100 µM (200, 300, 500, and 1,000 µM) were not recommended for *J. mimosaefolia* shoot multiplication because they caused partial browning of the explants, decreased the multiplication rate, and were not beneficial for the growth of shoots (data not presented). For further proliferation, the explants were subcultured on medium containing 100 μM KIN for 45 days.

Table 3. Effects of different cytokinins and their concentration on survival rate, shoot multiplication, and shoot growth of *J. mimosaefolia*

Cytok-inin	Concentration (μM)	Survival rate (%)	Shooting rate (%)	Shoot per explant Mean ± SE	Maximum length of shoots Mean ± SE
BAP	1	100	100	2.3 ± 0.37	2.4 ± 0.40
	10	100	100	3.4 ± 0.46	2.7 ± 0.48
	100	60	60	1.0 ± 0.00	1.6 ± 0.18
ZEA	1	100	100	1.8 ± 0.25	2.6 ± 0.45
	10	100	100	2.8 ± 0.31	2.2 ± 0.19
	100	0	0	---	---
KIN	1	100	100	1.0 ± 0.00	3.0 ± 0.15
	10	100	100	1.6 ± 0.32	2.3 ± 0.34
	100	100	100	6.1 ± 1.68	1.4 ± 0.32

SE: standard error.

Figure 2. Shoot multiplication of *J. mimosaefolia*.

Effects of Sucrose Concentrations

Four different concentrations of sucrose (0%, 1%, 2%, and 4% w/v) were tested to determine the effects of carbohydrate source on shoot multiplication. Shoot tips were cultured under a light intensity of 65 μmol m^{-2} s^{-1} for 45 days on B5 medium supplemented with 100 μM KIN (multiplication medium) containing each concentration of sucrose. Data were calculated from 10 explants for each concentration tested.

The results shown in Table 4 indicate that the concentration of sucrose added to the medium has a notable effect on the multiplication rate of shoot tips of *J. mimosaefolia*. The minimum number of proliferated shoots per explant (2.4) was achieved on sucrose-free medium. This rate was increased by about two-fold (5.2) on medium supplemented with 1% sucrose and was about three times greater (7.3) on medium supplemented with 2% sucrose. Supplementation of 4% sucrose did not increase the multiplication rate. The concentration of sucrose suitable for the multiplication of *J. mimosaefolia* under the conditions of this experiment was determined to be 2%; however, it is important to consider that the sucrose requirements depend on the stages of culture, sources of explant, and culture conditions. Differences in sucrose requirements have been reported among cultivars and even among explants from the same plant.

Effects of Gelling Agents

To determine the effects of gelling agents on shoot multiplication, shoot tips were cultured under a light intensity of about 65 μmol m^{-2} s^{-1} for 45 days on multiplication medium containing 2% (w/v) sucrose and solidified with 0.8% (w/v) Wako agar (Wako Pure Chemical Industries, Ltd., Osaka, Japan), 0.8% (w/v) Difco agar (Difco Laboratories, Detroit, USA), and 0.2% or 0.4% (w/v) Gelrite (Wako Pure Chemical Industries, Ltd., Osaka, Japan). Data were calculated from 10 explants for each treatment.

Table 5 shows the results of shoot multiplication after 45 days of culturing on media solidified with different gelling agents. The best multiplication rate was obtained on medium containing 0.2% Gelrite (8.2 shoots per explant), whereas the lowest number of shoots per explants was achieved with 0.8% Difco agar (3.8 shoots). Gelrite at a 0.4% level was not recommended because it resulted in a very firm medium, decreasing the multiplication rate.

Table 4. Effects of sucrose concentrations on shoot multiplication of *J. mimosaefolia*

Sucrose concentration (%)	Shoot per explant (Mean ± SE)
0	2.4 ± 0.43
1	5.2 ± 0.99
2	7.3 ± 1.78
4	5.8 ± 0.51

SE: standard error.

Table 5. Effects of gelling agents on shoot multiplication of *J. mimosaefolia*

Gelling agent	Concentration (%)	Shoot per explant (Mean ± SE)
Wako agar	0.8	6.8 ± 1.11
Difco agar	0.8	3.8 ± 0.57
Gelrite	0.2	8.2 ± 0.73
Gelrite	0.4	5.3 ± 0.97

SE: standard error.

Table 6. Effects of types of explants and light intensity on shoot multiplication of *J. mimosaefolia*

Type of explant	Light intensity (μmol ml^{-2} s^{-1})	Shoot per explant (Mean ± SE)
Apical shoot	"Low" (65)	8.5 ± 0.83
Apical shoot	"High" (135)	7.1 ± 0.80
Nodal segment	"Low" (65)	5.5 ± 0.45
Nodal segment	"High" (135)	3.3 ± 0.42

SE: standard error.

This result could be attributable to the difficulties faced by the cultured tissues to use the cytokinin and nutrients from the medium. The medium solidified with 0.8% Wako agar resulted in about a seven-fold propagation of shoots, which was less than that achieved using 0.2% Gelrite. Gelrite has been considered to be the most popular alternative to agar, although Gelrite was often reported to cause vitrification easily compared with agar.

However, in this experiment, no vitrification symptoms were observed in shoots cultured on medium containing 0.2%–0.4% Gelrite concentrations.

Effects of Types of Explants and Light Intensity

Generated shoots were isolated, cut into lengths of about 1.5 cm, classified into shoot-tip and nodal segment explants, and then cultured on multiplication medium containing 2% (w/v) sucrose and solidified with 0.2% (w/v) Gelrite, under two light intensities, i.e., "low" (65 μmol m^{-2} s^{-1}) and "high" (135 μmol m^{-2} s^{-1}). The effects on shoot multiplication were evaluated after 45 days of culture. Data were calculated from 10 explants for each treatment.

To improve the multiplication rate, propagation with nodal segment explants and propagation under a greater light intensity were attempted. However, there was no significant improvement in the multiplication rate after 45 days of culture (Table 6). The multiplication rate of nodal segment explants was significantly lower than that of shoot tips under both light intensity conditions tested. Light intensity "high" was not recommendable because it decreased the propagation rate in both shoot-tip and nodal segment explants.

Effects of Plant Density

To determine the effects of plant density on shoot multiplication, shoot tips (1–30 explants) were cultured in 300-ml Erlenmeyer flasks each containing 100 ml of multiplication medium supplemented with 2% (w/v) sucrose and solidified with 0.2% (w/v) Gelrite under a light intensity of 65 μmol m^{-2} s^{-1} for 45 days. The multiplication rate was determined by assessing the number of shoots per explant at different plant densities. Growth efficiency (number of new shoots per vessel) was calculated as (number of explants cultured per vessel × multiplication rate) − (number of explants cultured per vessel). Data were calculated from three replications for each plant density.

The results of shoot multiplication under different plant densities are shown in Table 7. The multiplication rate of *J. mimosaefolia* decreased with increasing plant densities of up to 10 shoots per vessel. However, a large difference in multiplication rate was not observed among plant densities of 1, 3, and 5 as well as between 10 and 15 shoots per vessel (300-ml Erlenmeyer flasks). Growth efficiency was found to be the highest at 15 shoots per culture vessel and decreased significantly at plant densities of 20 and 30 per vessel. This result indicates that the suitable density for shoot multiplication of *J. mimosaefolia* fluctuates between 10 and 15 shoots per culture vessel.

Table 7. Effects of plant density on shoot multiplication of *J. mimosaefolia*

Plant density (explant per vessel)	Multiplication rate Shoot per explant (Mean ± SE)	Growth efficiency New shoot per vessel (Mean ± SE)
1	15.0 ± 1.73	14.0 ± 1.73
3	14.7 ± 1.73	41.1 ± 5.20
5	14.0 ± 0.60	65.0 ± 3.00
10	8.5 ± 0.15	75.0 ± 1.52
15	7.2 ± 0.17	93.0 ± 2.60
20	4.4 ± 0.26	68.0 ± 5.29
30	2.7 ± 0.26	51.0 ± 7.94

SE: standard error.

Effects of Gelling Agents, Activated Charcoal (AC), and Light Intensity on Elongation of Proliferated Shoots

The effects of gelling agents, AC, and light intensity on the elongation of proliferated shoots were tested. Shoots were isolated, cut into lengths of 1.5 cm, and cultured under two light intensities, namely, "low" (65 μmol m^{-2} s^{-1}) and "high" (135 μmol m^{-2} s^{-1}) for 45 days on B5 medium supplemented with 1 μM KIN (growth medium) solidified with Wako agar or Gelrite and supplemented with 0%–0.5% (w/v) AC (Sigma Chemical

Co., St. Louis, USA; product number C6289). Data were calculated from 20 to 40 proliferated shoots for each treatment.

Table 8 shows the results of shoot elongation on media solidified with different gelling agents, with or without a supplement of AC and cultured under different light intensities. After 45 days of culture, the best shoot elongation was obtained on medium solidified with 0.2% Gelrite and supplemented with 0.5% AC. Similar to the effect on multiplication rate, Gelrite was better for shoot elongation than agar. Shoots on medium solidified with Gelrite grew 50%–70% better than those on medium solidified with agar. In addition, supplementation of AC was beneficial for shoot elongation, increasing the growth in about 15%-30% in comparison with the growth obtained on medium without AC. Activated charcoal often stimulates shoot elongation of several species (Bonga and Aderkas 1992). Presumably, AC absorbs harmful exudates such as phenolic oxidation products and other toxic components that have leached from the shoots into the medium after transfer to elongation medium. Like the result obtained in the experiment for shoot multiplication, the light intensity "high" did not improve shoot elongation. Although there were no significant differences in growth, the elongation of proliferated shoots was greatest under the light intensity "Low." For further elongation, the proliferated shoots were cultured under a light intensity of 65 μmol ml^{-2} s^{-1} on growth medium supplemented with 0.5% AC and 0.2% Gelrite (Figure 3).

Table 8. Effects of gelling agents, activated charcoal, and light intensity on elongation of proliferated shoots of *J. mimosaefolia*

Gelling agent (%)	Activated charcoal (%)	Light intensity (μmol m^{-2} s^{-1})	Length of shoots in cm (Mean ± SE)
Wako agar (0.8)	0	"Low" (65)	2.8 ± 0.27
Wako agar (0.8)	0	"High" (135)	2.6 ± 0.23
Wako agar (1.2)	0.5	"Low" (65)	3.7 ± 0.37
Wako agar (1.2)	0.5	"High" (135)	3.1 ± 0.29
Gelrite (0.2)	0	"Low" (65)	4.7 ± 0.18
Gelrite (0.2)	0.5	"Low" (65)	5.5 ± 0.26

SE: standard error.

Figure 3. Elongation of proliferated shoots of *J. mimosaefolia*.

Rooting of Proliferated Shoots

For rooting, half-strength B5 medium (1/2 B5) and half-strength MS medium (1/2 MS), supplemented with 0.49 or 4.90 μM indole-3-butyric acid (IBA) alone, or in combination with 0.27 μM naphthaleneacetic acid (NAA), were used (Table 9). Data were collected after 30 days of culture under a light intensity of 65 μmol m^{-2} s^{-1}, and rooting frequencies were calculated from 10 cultured shoots for each treatment.

Table 9. Effects of media and auxin concentration on the rooting of proliferated shoots of *J. mimosaefolia*

Medium	Auxin (μM)	Rooting rate (%)	Roots per explant	Maximum length of roots
			(Mean ± SE)	(Mean ± SE)
1/2 B5	IBA (0.49)	100	2.1 ± 0.35	0.9 ± 0.15
	IBA (0.49) + NAA (0.27)	100	3.0 ± 0.37	0.5 ± 0.11
	IBA (4.90)	100	3.1 ± 0.31	1.4 ± 0.9
	IBA (4.90) + NAA (0.27)	100	3.8 ± 0.20	0.9 ± 0.28
1/2 MS	IBA (0.49)	10	1.0 ± 0.00	0.3 ± 0.00
	IBA (0.49) + NAA (0.27)	40	3.0 ± 0.82	0.9 ± 0.27
	IBA (4.90)	60	2.3 ± 0.42	1.4 ± 0.34
	IBA (4.90) + NAA (0.27)	80	3.5 ± 0.98	1.3 ± 0.29

SE: standard error.

Results of the rooting experiments are shown in Table 9. Root initiation was observed within 1–2 weeks after transfer of shoots to rooting media. After 30 days of culturing, the frequency of rhizogenesis on 1/2 B5 medium was better than that on 1/2 MS. All shoots in all treatments on 1/2 B5 medium formed roots (Figure 4). On 1/2 MS medium, the best rooting percentage (80%) was achieved with 4.90 μM IBA + 0.27 μM NAA, and the worst result (10%) was achieved with 0.49 μM IBA (alone). The mean number of roots per explant obtained with combinations of IBA and NAA was greater than that obtained with IBA alone.

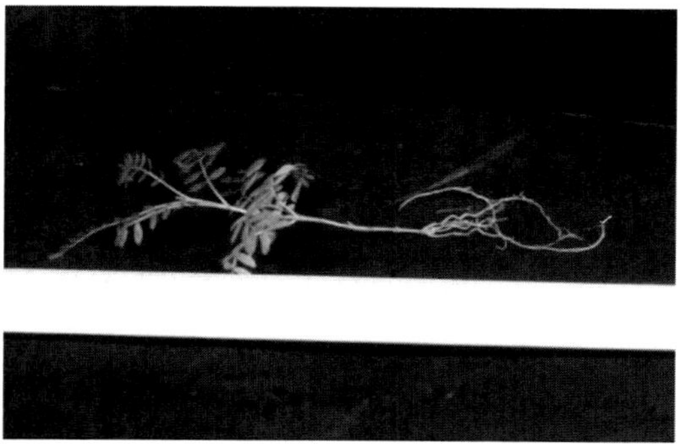

Figure 4. Rooting of proliferated shoot of *J. mimosaefolia*.

Development of *In vitro* Plantlets

For root development and further plantlet growth, rooted shoots were transferred to autoclaved 300-ml Erlenmeyer culture flasks containing 100 ml of vermiculite, sand, or perlite fertilized with 0.1% (v/v) Hyponex 5-10-5 plant food solution (The Hyponex Co., Inc., Osaka, Japan; containing in w/v: 5.00% N, 4.36% P, and 4.15% K) after the removal of adhering agar from roots with distilled water. Data were calculated from 10 plantlets for each substratum 45 days after transplanting.

Table 10 shows the results of survival and development of *in vitro* plantlets after 45 days of culturing in the culture room. Although the transfer of rooting shoots into vermiculite or sand substratum induced the development of healthy root systems and good growth of plantlets, the best result was obtained in flask containing vermiculite (Figure 5).

Table 10. Effects of different substrata on the survival and development of *in vitro* plantlets of *J. mimosaefolia*

Substratum	Survival rate (%)	Length of plantlets in cm (Mean ± SE)
Vermiculite	100	5.0 ± 0.72
Sand	100	3.6 ± 0.25
Perlite	60	2.7 ± 0.22

SE: standard error.

Figure 5. Plantlets of *J. mimosaefolia* growing *in vitro*.

Figure 6. Acclimatization of *J. mimosaefolia* plantlets.

Table 11. Survival and growth of *J. mimosaefolia* plantlets in the steps of acclimatization

Days after transplanting	Survival rate (%)	Length of plantlet in cm (Mean ± SE)
0	100	7.7 ± 0.26
45	100	13.0 ± 0.43
90	100	19.6 ± 0.29

SE: standard error.

Acclimatization of Plantlets

After 45 days of growth under culture room conditions, the plants were transplanted into plastic pots (10 cm diameter × 8 cm high) filled with vermiculite fertilized with 0.1% (v/v) Hyponex solution and acclimatized in a growth cabinet at 25°C–30°C under a light intensity of 35 μmol m^{-2} s^{-1} with a 16-h photoperiod. During the first 2 weeks, the plants were kept under high relative humidity in closed plastic boxes with transparent covers (Figure 6). Covers were opened gradually during the next 2 weeks, covering about 99% of plastic boxes during the first 5 days and then about 97%, 90%, and 80% during 3 days for each condition. Thereafter, the covers were removed. Data were calculated from 10 plants 0, 45, and 90 days after transplanting.

Figure 7. Acclimatized plantlet of *J. mimosaefolia*.

Developed plantlets were transplanted into plastic pots containing vermiculite and acclimatized successfully under the above mentioned conditions (Figure 7). Maintaining a high relative humidity for at least 14 days was required for high survival rates. Similar results have been reported for plant acclimatization of *Betula pendula* (Iliev et al. 2001), *Cedrela odorata* (Maruyama et al. 1989), and *Guazuma crinita* (Maruyama et al. 1996). Afterward, the plants were planted in the greenhouse. In another experiment, successful acclimatization of plantlets was achieved by transplanting the rooted shoots directly into pots placed in the growth cabinet, omitting the previous step for root development and growth of shoots in flasks kept under culture room conditions. Ninety days after transplanting, all plants survived and reached 20 cm height in average (Table 11). This method requires less handling and may reduce the period necessary for outplanting.

Alginate-Encapsulated Technology for Propagation and Germplasm Conservation

Construction of Artificial Seeds

For the construction of single-layered beads, shoot-tip and nodal segment explants (including one axillary bud), measuring about 3–4 mm long, were aseptically excised from *in vitro* plantlets and then immersed into autoclaved B5 medium containing 1 μM KIN and 4% (w/v) sodium alginate. Next, the explants mixed with the alginate medium were picked up using tweezers and dropped into a sterile solution of 1.4% (w/v) calcium chloride where they remained for 30 min. After that, the calcium chloride solution was decanted and the constructed alginate beads were rinsed three times with autoclaved medium (Figure 8).

Double-layered beads were constructed by the method described by Kinoshita and Saito (1992). As shown in Figure 8, shoot-tip explants and calcium alginate single-layered beads constructed previously were mixed with a 3% (w/v) sodium alginate medium. Each explant, together with a single-layered bead mixed with sodium alginate medium, was picked up using tweezers and dropped into calcium chloride solution again. The explant was contained in the outer layer.

Regeneration of Plantlets under Aseptic Conditions

Regeneration of plantlets from encapsulated shoot-tip or nodal segment explants was tested under aseptic conditions. Artificial seeds were cultured under a photon flux density of about 65 μmol m^{-2} s^{-1} with a light regimen of 16 h daily in a culture room at 25°C.

Effect of Type of Encapsulated Explants and Substrata on Bud Emergence and Shoot Growth of Artificial Seeds

Two types of explants, shoot-tip and nodal segment, were encapsulated into single-layered alginate gel beads and cultured *in vitro* on different substrata, namely, water or medium solidified with 1% (w/v) agar with or without plant growth regulators or sucrose.

Table 12. Effect of encapsulated explant type and substratum on the bud emergence and shoot growth of artificial seeds of *J. mimosaefolia* under aseptic conditions

Substratum	Shoot-tip		Nodal segment	
	Bud emergence (%)	Shoot growth (%)	Bud emergence (%)	Shoot growth (%)
W	30	0	0	0
W + H	40	0	0	0
W + S	100	100	60	20
W + S + H	100	100	50	20
M	20	20	0	0
M + H	40	20	0	0
M + S	100	100	80	40
M + S + H	100	100	100	100

W: water, H: 1 μM KIN, S: 2% (w/v) sucrose, M: sucrose and plant growth regulator-free B5 medium

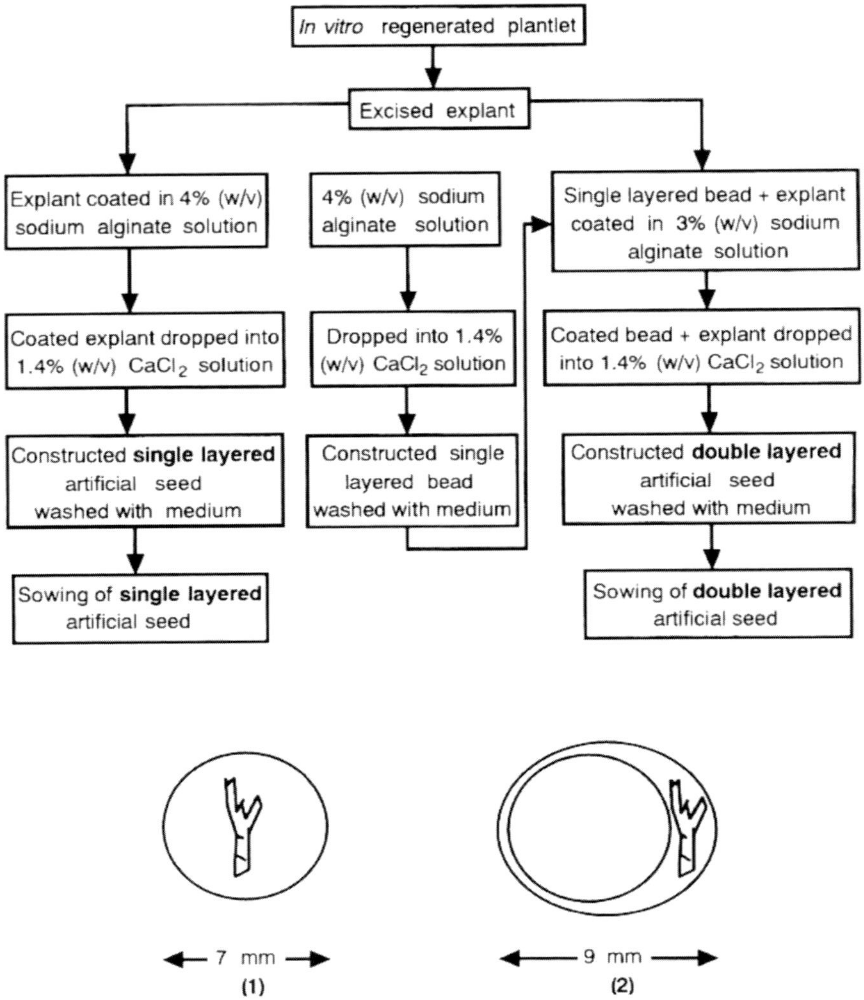

Figure 8. Construction schemes of artificial seeds.

Table 12 shows the results of bud emergence and shoot growth of the two types of encapsulated explants on different substrata. When encapsulated explants of both shoot-tip and nodal segment were cultured on a medium substratum supplemented with sucrose, high percentages of bud emergence and shoot growth were obtained. This result suggests that sucrose is essential to promote bud emergence and shoot growth of *J.*

mimosaefolia. Although no differences were observed when medium supplemented with sucrose and plant growth regulators was used, in general, shoot-tip explants responded better than nodal segment explants.

Effect of Concentration of Medium Contained in the Gel Bead Matrix on the Bud Emergence and Shoot Growth of Artificial Seeds

Shoot-tip explants were encapsulated into single-layered alginate gel beads containing medium at different concentrations (100%–2000%) and cultured on water substratum solidified with 1% (w/v) agar. Medium at a concentration of 100% represents the standard basal medium. Medium at 500%, 1000%, and 2000%, represents 5, 10, and 20 times the concentration (w/v) of the standard basal medium, respectively, excepting the concentration of calcium components of the medium, which was added at standard concentrations in all treatments. Results shown in Table 13 indicate that artificial seeds showed increased bud emergence and shoot growth rates proportionately to the concentration of medium contained in the gel bead matrix. Similar results have been reported for artificial seeds of the Japanese white birch (Kinoshita and Saito 1990).

Figure 9. Plant regeneration from artificial seed of *J. mimosaefolia* under aseptic conditions.

Table 13. Effect of medium concentration contained in gel bead matrix on the bud emergence and shoot growth of artificial seeds of *J. mimosaefolia* under aseptic conditions

Concentration of medium (%)	Bud emergence (%)	Shoot growth (%)
100	30	0
500	50	0
1000	60	30
2000	70	50

Concentration of 100% medium represents the standard basal medium.

Effect of Type of Beads on the Bud Emergence and Shoot Growth of Artificial Seeds

In an attempt to improve the plant regeneration rate from artificial seeds on substratum without nutrients, shoot-tip explants encapsulated into four types of beads were tested. Four types of beads (A, B, C, and D, Table 14) were made from combinations of single or double layers and two concentrations of medium. They were cultured on water substratum solidified with 1% (w/v) agar under culture room conditions.

Table 14. Effect of type of bead on the bud emergence, shoot growth, and rooting of artificial seeds of *J. mimosaefolia* under aseptic conditions

Type of bead	Bud emergence (%)	Shoot growth (%)	Rooting (%)
A	30	0	0
B	60	30	0
C	100	100	0
D	100	100	60

A: single-layered beads containing medium at a concentration of 100%, B: single-layered beads containing medium at a concentration of 1000%, C: double-layered beads containing medium at a concentration of 1000% in the inner layer and a concentration of 100% in the outer layer, D: double-layered beads containing medium at a concentration of 1000% supplemented with 0.5% (w/v) activated charcoal in the inner layer and a concentration of 100% in the outer layer.

The results obtained after 45 days of culturing, shown in Table 14, indicate that double-layered beads were much better than single-layered beads in stimulating bud emergence, shoot growth, and rooting of artificial seeds. Moreover, addition of 0.5% (w/v) activated charcoal into the gel matrix of the inner layer significantly enhanced root development (Figure 9) and rooting rate of *J. mimosaefolia* (Table 14).

Regeneration of Plantlets under Non-Aseptic Conditions

Shoot-tip explants encapsulated into different types of beads were placed in plastic pots containing non-sterilized perlite substratum saturated with water and initially covered with a Petri dish and sealed with parafilm to prevent drying. Plastic pots containing artificial seeds were kept in a growth cabinet at 25°C–30°C under a 16-h photoperiod with a photon flux density of about 35 $\mu mol\ m^{-2}\ s^{-1}$ provided by cool fluorescent lamps. They were irrigated with water until the evident appearance of roots from grown shoots and then fertilized with 0.1% (w/v) Hyponex 5-10-5 plant food solution.

Table 15. Effect of type of bead on the bud emergence, shoot growth, and rooting of artificial seeds of *J. mimosaefolia* under non-aseptic conditions

Type of bead	Bud emergence (%)	Shoot growth (%)	Rooting (%)
B	12.5	0	0
C	86.7	26.7	26.7
D	85.0	31.7	31.7

B: single-layered beads containing medium at a concentration of 1000%, C: double-layered beads containing medium at a concentration of 1000% in the inner layer and a concentration of 100% in the outer layer, D: double-layered beads containing medium at a concentration of 1000% supplemented with 0.5% (w/v) activated charcoal in the inner layer and a concentration of 100% in the outer layer.

Table 15 shows the results of plant regeneration from encapsulated shoot tips under non-aseptic conditions. When single-layered artificial seeds were sowed directly into non-sterilized perlite substratum, the bud emergence rate was 12.5% but without subsequent shoot growth and rooting. In contrast, double-layered artificial seeds showed bud emergence and plant regeneration rates of about 85% and 30%, respectively (Table 15, Figures 10 and 11).

Figure 10. Bud emergence of artificial seeds of *J. mimosaefolia* under non-aseptic conditions.

Figure 11. Plant regeneration of artificial seeds of *J. mimosaefolia* under non-aseptic conditions.

Storage of Artificial Seeds

Encapsulated shoot tips were transferred to 200-ml Erlenmeyer flasks containing 75 ml of the following substrates: water, water containing 2%

(w/v) sucrose, and sucrose and plant growth regulator-free B5 medium. Substrates were solidified with 1% (w/v) agar. Storage was continued for 3, 6, and 12 months at 25°C and 20°C (16-h photoperiod under a photon flux density of about 35 μmol m^{-2} s^{-1}) and at 12°C (in the dark). The stored encapsulated shoot tips were transferred to germination medium and incubated in a culture room at 25°C under a photon flux density of about 35 μmol m^{-2} s^{-1} for 16 h daily.

Table 16. Effect of substrates and storage temperatures on plant conversion from stored alginate-encapsulated shoot tips of *J. mimosaefolia*

Months of storage	25°C★			20°C			12°C		
	W	W + S	M	W	W + S	M	W	W + S	M
3	40	90	30	90	20	0	70	70	0
6	0	0	0	70	10	0	0	80	0
12	0	0	0	0	0	0	0	10	0

W: water, S: 2% (w/v) sucrose, M: sucrose and plant growth regulator-free B5 medium, ★: Germination occurred during storage.

Table 16 shows the results of plant conversion from alginate-encapsulated shoot tips after 3, 6, and 12 months of storage under various conditions. Storage of beads at 25°C was not possible because of germination during storage.

The best temperature for storage of encapsulated shoot tips was 20°C on water substratum or 12°C on water substratum containing 2% (w/v) sucrose. Shoot tips encapsulated in alginate beads were stored for at least 6 months without considerable loss of viability. Storage of alginate-encapsulated shoot tips at temperatures above freezing was found to be an acceptable method for medium-term conservation of *J. mimosaefolia* germplasm cultured *in vitro*.

CONCLUSION

The possibility of rapid clonal production by micropropagation of *J. mimosaefolia* was proven by the culturing of shoot tips. In addition, although further studies (including development of suitable coating for artificial seeds that permit easy plant conversion under non-aseptic conditions) are necessary to improve the efficiency of the alginate-encapsulated system, the method described here can be an attractive possibility for the propagation and germplasm conservation of Jacaranda at temperature conditions above freezing. Storage of alginate-encapsulated germplasm at room temperature can be used as a good alternative for the conservation of several tropical species that are intolerant to conventional reduced-temperature storage systems.

REFERENCES

Bajaj, Y. P. S. (1989). Some Indian ornamental trees: *Cassia fistula* Linn., *Ponciana regia* (Boj.) and *Jacaranda acutifolia* Auct *In* Y.P.S. Bajaj (Eds.), *Biotechnology in Agriculture and Forestry. Trees II*, vol. 5, (pp. 469-476), Springer-Verlag.

Bonga, J. M., and von Aderkas, P. (1992). *In vitro cultures of trees*. Kluwer (pp. 236), Academic Publishers.

Chalupa, V. (1981). Clonal propagation of broad-leaved forest trees *in vitro*. *Communicationes Instituti Forestalis Cechosloveniae, 12,* 255-271.

Debergh, P. C., and Read, P. E. (1991). Micropropagation. In P. C. Debergh and R. H. Zimmerman (Eds.), *Micropropagation Technology and Application* (pp. 1-13). Kluwer Academic Publishers.

Gamborg, O. L., Miller, R. A., and Ojima, K. (1968). Nutrient requirements of suspension cultures of soybean root cells. *Experimental Cell Research*, 50, 151-158.

Holdgate, D. P. (1977). Propagation of ornamentals by tissue culture. In J. Reinert and Y. S. P. Bajaj (Eds.), *Applied and Fundamental Aspects of Plant Cell, Tissue and Organ Culture* (pp. 18-42). Springer-Verlag.

Iliev, I., Kitin, P., and Funada, R. (2001). Morphological and anatomical study on *in vitro* root formation of Silver Birch (*Betula pendula* Roth.). *Propagation of Ornamental Plants, 1,* 10-19.

Kinoshita, I., and Saito, A. (1990). Propagation of Japanese white birch by encapsulated axillary buds (1) Regeneration of plantlets under aseptic conditions. *Journal of the Japanese Forest Society, 72,* 166-170.

Kinoshita, I., and Saito, A. (1992). Regeneration of Japanese white birch plants from encapsulated axillary buds (pp. 493-496). In M. Kuwahara and M. Shimada (Eds.), *5th International Conference on Biotechnology in the Pulp and Paper Industry* (pp. 493-496)., Kyoto-Japan, Uni Publishers: Tokyo.

Lloyd, G., and McCown B. (1981). Commercially-feasible micro-propagation of Mountain Laurel, *Kalmia latifolia* by use of shoot tip culture. *International Plant Propagator's Society Combined Proceedings for 1980, 30,* 421-427.

Maruyama, E., Ishii, K., Saito, A., and Migita, K. (1989). Micro-propagation of Cedro (*Cedrela odorata* L.) by shoot-tip culture. *Journal of the Japanese Forest Society, 71,* 329-331.

Maruyama, E., Ishii, K., Saito, A., and Ohba, K. (1993). Micropropagation of Jacaranda (*Jacaranda mimosaefolia* D. Don) by shoot-tip culture. *Journal of the Japanese Forest Society, 75,* 346-349.

Maruyama, E., Ishii, K., Kinoshita, I., Ohba, K., and Saito, A. (1996). Micropropagation of Bolaina Blanca (*Guazuma crinita* Mart.), a fast-growing tree in the Amazon region. *Journal of Forestry Research, 1,* 211-217.

Maruyama, E., Kinoshita, I., Ishii, K., Ohba, K., and Saito, A. (1997a). Germplasm conservation of the tropical forest trees, *Cedrela odorata* L., *Guazuma crinita* Mart., and *Jacaranda mimosaefolia* D. Don., by shoot tip encapsulation in calcium-alginate and storage at 12-25°C. *Plant Cell Reports, 16,* 393-396.

Maruyama, E., Kinoshita, I., Ishii, K., Shigenaga, H., Ohba, K., and Saito, A. (1997b). Alginate-encapsulated technology for the propagation of the tropical forest trees: *Cedrela odorata* L., *Guazuma crinita* Mart., and *Jacaranda mimosaefolia* D. Don. *Silva Genetica, 46,* 17-23.

Maruyama, E. (2003). *In vitro* propagation of Jacaranda (*Jacaranda mimosaefolia* D. Don). *Propagation of Ornamental Plants, 3,* 27-33.

Murashige, T. (1974). Plant propagation through tissue culture. *Annual Review of Plant Physiology, 25,* 135-165.

Murashige, T., and Skoog, F. (1962). A revised medium for rapid growth and bioassays with tobacco tissue cultures. *Physiologia Plantarum, 15,* 473-497.

Thorpe, T. A., Harry, I. S., and Kumar, P. P. (1991). Application of micropropagation to forestry. In P. C. Debergh, R. H. Zimmerman (Eds.), *Micropropagation Technology and Application* (pp. 311-336). Kluwer Academic Publishers.

In: Micropropagation of Ornamental Plants ISBN: 978-1-53614-541-0
Editor: Taoufik Bettaieb

Chapter 7

PROPAGATION OF RYUKYUMATSU (*PINUS LUCHUENSIS* MAYR.) VIA SOMATIC EMBRYOGENESIS

Yoshihisa Hosoi* and Tsuyoshi E. Maruyama
Forestry and Forest Products Research Institute (FFPRI)
Department of Forest Molecular Genetics and Biotechnology
Tsukuba, Ibaraki, Japan

ABSTRACT

Somatic embryogenesis in Ryukyumatsu (*Pinus luchuensis* Mayr.), an endemic species in Ryukyu Island, Japan, was initiated from megagametophytes containing zygotic embryos on a medium supplemented with 10 μ*M* 2,4-dichlorophenoxyacetic acid and 5 μ*M* 6-benzylaminopurine. Embryogenic cultures were maintained and proliferated by subcultures at 2- to 3-weeks intervals on the same fresh medium. The maturation of somatic embryos occurred on media containing maltose, activated charcoal, abscisic acid and polyethylene glycol. High frequencies of germination were obtained after the post-maturation treatment of somatic

---

* Corresponding Author Email: yh2884@ffpri.affrc.go.jp.

embryos under conditions of high relative humidity, and around 90% of them were converted into plants. Furthermore, a procedure for protoplast isolation from embryogenic tissue was also described.

Keywords: embryo culture, micropropagation, Ryukyu Island, *Pinus luchuensis*, polyethylene glycol, protoplast culture, Ryukyu pine, somatic embryogenesis

INTRODUCTION

Ryukyumatsu (*Pinus luchuensis* Mayr.), a two-needle pine species endemic to Nansei-shoto (south-western islands of Kyushu and in the Okinawan archipelago), is the Japanese representative of a group of three closely related taxa, including *P. taiwanensis* of Taiwan and *P. hwangshanensis* of mainland China. Ryukyumatsu is also one of the most economically important trees in Okinawa Prefecture and is valued for the construction of homes and furniture, the production of decorative veneer, and for landscaping (Figure 1) (Hosoi and Maruyama, 2012). However, in recent years, Ryukyumatsu populations have further declined as a result of pine wilt disease, caused by the pinewood nematode, *Bursaphelenchus xylophilus* (Steiner et Buhrer) Nickle, which is primarily transmitted by the Japanese pine sawyer, *Monochamus alternatus* Hope (Akiba and Nakamura, 2005). The pinewood nematode and pine sawyer are among the most serious pests in Japan, while pine wilt disease has been a key critical factor in the mass mortality, not only of the Ryukyumatsu populations but also in other important Japanese pine forests consisting of *P. densiflora*, *P. thunberghii* or *P. armandii* var. *amamiana* (Maruyama and Hosoi, 2014). The nematode is inferred to be native to North America and since its introduction into Japan at the beginning of the 20th century, the pinewood nematode has spread to Korea, Taiwan, and China and has devastated pine forests in East Asia (Togashi and Shigesada, 2006), and it was also found in Portugal in 1999 (Mota et al. 1999).

Somatic embryogenesis is one of the most promising techniques for mass propagation of selected resistant trees, for the *ex situ* conservation of genetic resources by cryopreservation techniques, and for plant regeneration in genetic transformation protocols (Park et al. 1998, Bonga, 2016, Maruyama and Hosoi, 2016). In this chapter, we describe the initiation of embryogenic cultures from seed explants, production of somatic embryos, regeneration of plants, and protoplast isolation from embryogenic tissues.

EMBRYOGENIC CULTURE INITIATION AND PROLIFERATION

Open-pollinated cones were collected from a mother tree at the Iriomote Tropical Forest Tree Breeding Technical Garden (Taketomi, Okinawa, Japan). Collected cones were disinfected by 5 min immersion in 99.5% ethanol, and then dried in a laminar-flow cabinet before dissection. Excised seeds were disinfected with 2.5% (v/v available chlorine) sodium hypochlorite solution for 30 min, and then rinsed 5 times with sterile distilled water. After the seed coats had been removed, the megagametophytes (containing zygotic embryos at the cotyledonary developmental stage) were used as explants for the initiation of embryogenic cultures.

The explants were cultured in 4-compartment plates (100 × 15 mm) (Kord-Valmark Labware, Ontario, Canada) containing somatic embryogenesis induction medium. The induction medium was EM medium (Maruyama et al. 2000), modified as follows: basal salts, vitamins, and myo-inositol were reduced to half the standard concentrations; the concentrations of KCl was reduced to 40 mg l^{-1}; and 500 mg l^{-1} casein hydrolysate as well as 1 g l^{-1} L-glutamine, 10 g l^{-1} sucrose, 10 μ*M* 2,4-D, and 5 μ*M* BA were added. The medium was solidified with 3 g l^{-1} gellan gum (Gelrite®; Wako Pure Chemical, Osaka, Japan).

The pH of the medium was adjusted to 5.8 prior to autoclaving, and the cultures were kept in darkness at 25°C. The presence or absence of somatic embryos at distinct early stages characterized by an embryo part (smaller and denser cells) with a suspensor (long vacuolated cells) was determined weekly under an inverted microscope up to 12 weeks. Induced embryogenic tissues were transferred onto the same fresh medium supplemented with 30 g l^{-1} sucrose, 3 μ*M* 2,4-D, and 1 μ*M* BA, but lacking casein hydrolysate. Once embryogenic cultures increased in mass, they were maintained and proliferated by subculturing 10−20 pieces of tissue per plate at 2- to 3- week-intervals and incubating under conditions equivalent to those for embryogenic culture initiation.

Figure 1. *Pinus luchuensis* using as ornamental trees.

The extrusion of embryogenic tissue from the micropylar end of explants occurred after about 3−4 weeks of culture (Figure 2A). An average of 1% (2/192) of megagametophytes cultured extruded embryogenic tissues, which were then removed and transferred to a maintenance-proliferation medium. However, after several subcultures, only one could be proliferated and converted into stable embryogenic line. Subsequently, embryonic masses containing cells characterized by dense cytoplasm cells (embryonic heads) and elongated cells (suspensor systems), proliferated readily by subculturing at 2- to 3- week-intervals on a medium supplemented with 3 μ*M* 2,4-D and 1 μ*M* BA, retaining their original translucent and mucilaginous appearance (Figure 2B). The FW of tissue on the maintenance-proliferation medium increased 5- to 10-fold after a 2- to 3- week-culture period (data not shown).

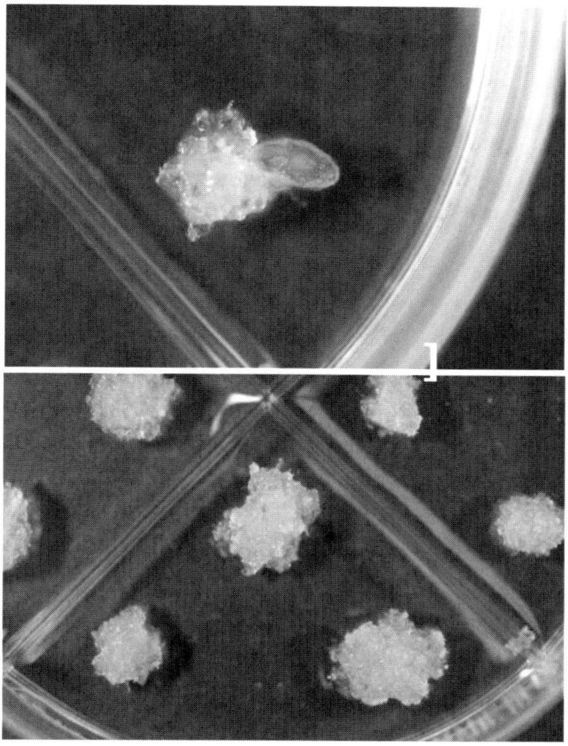

Figure 2. Somatic embryogenesis initiation (A) and proliferation of embryogenic tissues (B) in *Pinus luchuensis.*

MATURATION OF SOMATIC EMBRYOS

Embryogenic tissues, about 2 weeks after subculture, were used for maturation experiments. Five pieces of embryogenic tissues (each about 200 mg in FW) were transferred to 90 x 20 mm monoplates (Iwaki, Asahi Glass Co., Ltd, Tokyo, Japan) containing 30–40 ml of semi-solid maturation medium. The maturation medium contained salts and vitamins from the original EM medium (Maruyama et al. 2000), 30 g l–1 maltose,

100 μM ABA, 0–200 g l–1 PEG 6000 (Av. Mol. Wt.: 7,300–9,300; Wako Pure Chemical, Osaka, Japan), 2 g l–1 AC (Wako Pure Chemical, Osaka, Japan), 3 g l^{-1} gellan gum, and amino acids (g l^{-1}: glutamine 0.75, asparagine 0.5, arginine 0.25, citrulline 0.04, ornithine 0.04, lysine 0.03, alanine 0.02 and proline 0.02). The pH of the medium was adjusted to 5.8 prior to autoclaving. Plates were sealed with Parafilm and kept in darkness at 25°C. Mature somatic embryos represent cotyledonary embryos with an elongated embryonic region and well-developed cotyledons, resembling zygotic embryos at the late cotyledonary stage.

Somatic embryo development and maturation were promoted on maturation medium supplemented with 30 g l^{-1} maltose, 100 μ*M* ABA, 2 g l^{-1} AC and 100–200 g l^{-1} PEG. About 2 weeks after transfer onto maturation medium, embryogenic cells developed gradually to form an individual and compact mass (pro-embryo). Cotyledonary somatic embryos were first observed about 4 weeks after the transfer of the embryonic mass, and were clearly distinct after 6 weeks of culture (Figure 3).

Figure 3. Cotyledonary somatic embryos of *Pinus luchuensis*.

The PEG in the medium stimulated the embryo maturation, and the number of mature embryos increased with increased PEG concentration up to 150 g l^{-1}, but decreased at 200 g l^{-1} (Figure 4). Although the PEG concentration from 100 to 200 g l^{-1} did not result in any statistical difference in terms of the somatic embryo yield per plate, the highest embryo maturation frequency was obtained on a medium supplemented with 150 g l^{-1}, with an average of 282 cotyledonary embryos collected per plate, in comparison with 244 and 228 embryos per plate at concentrations of 100 and 200 g l^{-1}, respectively (Figure 4). In the absence of PEG, the embryogenic cell proliferation was prominent and only a few early somatic embryos developed into the cotyledonary stage (average of 11 embryos per plate).

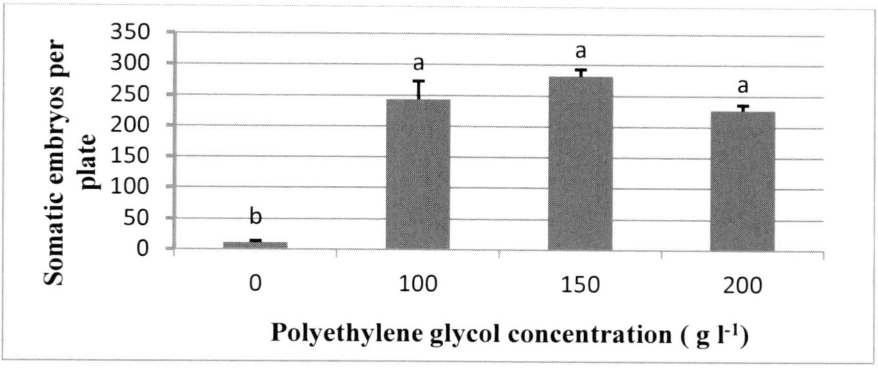

Figure 4. Effect of polyethylene glycol concentration on maturation of *Pinus luchuensis* somatic embryos. Values represent the mean of five plates for each concentration. Bars indicate the standard error. Means followed by same letter are not significantly different at $p < 0.05$.

Somatic Embryo Germination, Plantlet Conversion, and Acclimatization of Somatic Plants

Mature somatic embryos collected from the maturation medium after 8 weeks of culture were subjected to slow desiccation at high relative humidity prior to germination. Somatic embryos were placed on filter paper sheets (about 40 × 30 mm in size) equipped in 2 (central) wells

(Figure 5) of a 6-well multiplate (Iwaki, Asahi Glass Co., Ltd, Tokyo, Japan) in which the remaining 4 (side) wells were filled with 5–6 ml of sterile water, sealed tightly with Parafilm, and kept in darkness at 25°C for 3 weeks.

Under these conditions, the generated relative humidity inside the plate was approximately 98%, which was registered with a thermo-hygrometer recorder (RS-10, ESPEC MIC Corp. Aichi, Japan). After post-maturation treatment, somatic embryos were transferred to 6 g l^{-1} gellan gum-solidified germination medium, which composition was equivalent to that for the maintenance and proliferation, except for the elimination of plant growth regulators and supplementation with 30 g l^{-1} glucose, 2 g l^{-1} AC, 0.4 g l^{-1} glutamine, 0.25 g l^{-1} arginine, and 0.1 g l^{-1} proline. Cultures were kept at 25°C under a photon flux density of about 65 μmol m^{-2} s^{-1} provided by cool, white fluorescent lamps (100V, 40W; Toshiba, Tokyo, Japan) for 16 h. The numbers of somatic embryos that germinated (root emergence) and grew into plantlets (development of epicotyls from germinated embryos) were recorded after 6 and 12 weeks, respectively.

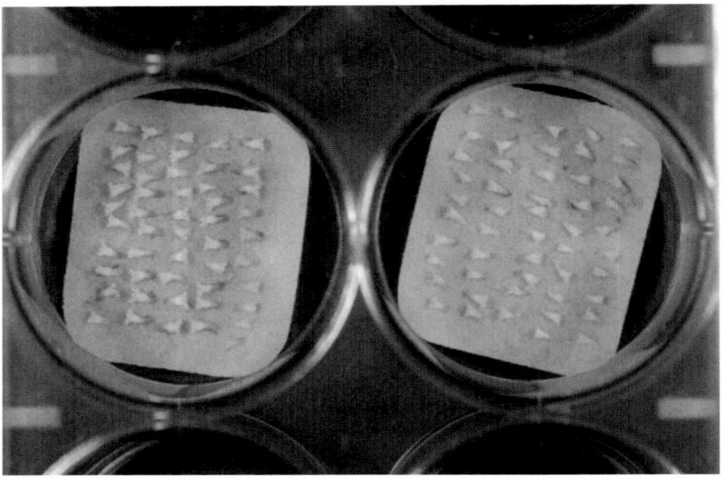

Figure 5. Desiccation of *Pinus luchuensis* somatic embryos.

Table 1. Germination and conversion frequencies in somatic embryos of *Pinus luchuensis* after maturation on media with different concentrations of polyethylene glycol (PEG)

Somatic embryos from maturation media containing PEG (g l^{-1})	Frequency of germination (% ± SE)	Frequency of conversion (% ± SE)
0	NT	NT
100	90 ± 1.2	88 ± 2.1
150	90 ± 4.1	87 ± 6.0
200	85 ± 2.4	82 ± 1.8

Germination and conversion frequencies represent the mean ± SE (standard error), n = 5 replicates of 30 to 50 somatic embryos per plate. NT: no tested

The regenerated plantlets were transferred into 300-ml flasks containing 100 ml of fresh germination medium with no amino acids and containing 30 g l^{-1} sucrose or glucose, 5 g l^{-1} AC and 10 g l^{-1} agar (Wako Pure Chemical Industries, Osaka, Japan) or into Magenta® vessels (Sigma, St. Louis, USA) containing Florialite® (Nisshinbo Industries, Tokyo, Japan) irrigated with a plant food solution modified from Nagao (1983) (containing in mg l^{-1}: NH_4NO_3 143, $NaH_2PO_4 \cdot 2H_2O$ 55.1, KCl 47.1, $CaCl_2 \cdot 2H_2O$ 52.5, $MgSO_4 \cdot 7H_2O$ 61, Fe-III EDTA 25, Cu EDTA 0.1, Mn EDTA 0.1, Zinc EDTA 0.1, H_3BO_3 1.5, KI 0.01, $CoCl_2 \cdot 6H_2O$ 0.005, and MoO_3 0.005), and kept under conditions equivalent to those described above for about 20 weeks prior to *ex vitro* acclimatization.

Table 1 shows the germination and conversion frequencies after post-maturation treatment (Figure 5) of somatic embryos that matured on media with different concentrations of PEG. Radicle emergence was observed to commence 1—2 weeks after transfer to the germination medium, and after 6 weeks of culture, the germination (Figure 6) frequencies varied from 85 to 90%. Subsequently, epicotyls developed from germinants, and after 12 weeks of culture, the plantlet conversion frequencies ranged from 82 to 88% (Table 1). Although the germination and conversion frequencies decreased slightly with increasing PEG concentration in maturation media, no morphological differences in germinants and plantlets were observed among PEG-derived treatments.

Regenerated plants (Figure 7) were cultured *in vitro* for about 20 weeks before *ex vitrum* acclimatization. The survival rate and the growth of plantlets after 15 weeks of culture are shown in Table 2. Survival rates of 100% in all treatments were achieved, and the optimal result in term of growth was obtained using Magenta® vessels containing Florialite® substrate irrigated with Nagao's plant food solution (Figure 8). For growth on semi-solid media, at the concentration of sugar tested, the growth of plantlets improved when supplementing the germination medium with sucrose in lieu of glucose.

Figure 6. Germination of *Pinus luchuensis* somatic embryos.

Figure 7. Regenerated *Pinus luchuensis* somatic plants.

Table 2. Effect of kind of media/substrate on survival and growth of *Pinus luchuensis* somatic plants

Media/Substrate	Survival (% ± SE)	Growth (cm ± SE)
Medium with sucrose	100 ± 0.0	3.5 ± 0.2
Medium with glucose	100 ± 0.0	2.9 ± 0.1
Fluorialite	100 ± 0.0	4.4 ± 0.2

Survival and growth of somatic plants were evaluated after 15 weeks of culture on germination media containing 30 g l^{-1} sucrose or glucose, and into Magenta® vessels containing Florialite® irrigated with Nagao's plant food solution. Values were calculated from 30 somatic plants for each treatment. SE: standard error

Subsequently, developed plantlets were acclimatized successfully under the conditions described as follows (Figure 9) (data not shown). The developed plants were transplanted into plastic pots filled with Kanuma soil and acclimatized inside a growth chamber at 25°C and 80% relative humidity. During the first 2 weeks, the somatic embryo-derived plants were kept in plastic boxes with transparent covers. Subsequently, the covers were gradually opened during a further 2- week-period and removed completely about 1 month after transplanting. The plants were irrigated with tap water for the first 2 weeks and then with a 0.1% (v/v) Hyponex® 6—10—5 plant-food solution (Hyponex Japan Co., Ltd., Osaka, Japan).

PROTOPLAST ISOLATION AND CULTURE

Embryogenic tissues, about 2 weeks after subculture, were used for protoplast isolation. About 7 g FW of tissue was transferred to 40 ml of an enzymatic mixture containing 1% Cellulase Onozuka RS (Yakult Pharmaceutical Industry Co., Ltd., Tokyo, Japan), 0.2% Pectolyase Y-23 (Kyowa Hakko Kirin Co., Ltd., Tokyo, Japan), and 0.6M mannitol. The enzymatic mixture containing embryogenic tissue in a 100 ml culture flask was incubated for 6 h at 25°C in darkness. The resulting protoplast suspension was diluted with 0.6 M mannitol solution and passed through a 30 μm polyester screen. After 3 min centrifugation at 60 × g, the sedimented protoplasts were washed twice by resuspending in 0.6 M mannitol solution and subsequent sedimentation. The yield of the protoplast was determined with a hemacytometer and their viability assessed using FDA. Protoplasts were cultured at a density of 2×10^3 ml^{-1} into a 96-well multiple plate containing 50 μl of liquid 1/2 EM media, which, in turn, contained 0.6 M mannitol supplemented with various combinations of 2,4-D (0.1, 0.3, 1, 3, 10, 30 μM) and BA (0, 1, 3, 10 μM). Plates containing protoplast culture in wells and sterile distilled water in outspaces among the wells to maintain relative humidity, were sealed with Parafilm and kept at 25°C in darkness.

Figure 8. Regenerated *Pinus luchuensis* somatic plants growing in Magenta® vessels containing Florialite® substrate irrigated with Nagao's plant food solution.

Preliminary experiments (data not shown) testing different combinations of enzymes, concentrations of mannitol, and combinations of plant growth regulators determined conditions for the isolation and culture of protoplasts from embryogenic tissues (Figure 10A). The viability examination by FDA staining showed that the survival rate of protoplasts ranged from 96 to 100%. This result indicated that an enzyme solution containing 1% Cellulase Onozuka RS, 0.2% Pectolyase Y-23, and 0.6 M mannitol was effective for protoplast isolation from embryogenic material.

Figure 9. Acclimatized *Pinus luchuensis* somatic plants.

The protoplast yield achieved was 1.32-1.46 × 10^5 protoplasts/g FW, and the size of isolated protoplasts varied mostly from about 20 to 60 μm in diameter (Figure 10B). The result in terms of protoplast yield achieved in our experiments for Ryukyumatsu was consistent to the reported yields from embryogenic tissue of *Pinus caribaea* (6.8-9.7 × 10^4 protoplasts/g FW, David et al. 1995), *Picea mariana* (0.5-4 × 10^5 protoplasts/g FW, Tautorus et al. 1990), and *Picea glauca* (1.2-4.8 × 10^5 protoplasts/g FW, Attree et al. 1989). Cell division and colony formation was mostly observed after 3-5 and 7-10 days of culture, respectively. Colony formation was achieved for all 24 combinations of plant growth regulators, but efficiency peaked at a concentration of 3-10 μ*M* 2,4-D plus 1-3 μ*M* BA (data not shown). The patterns of cell division (Figure 10C) and the subsequent colony formation (Figure 10D) observed were similar to those reported for the Japanese black pine (Hosoi and Maruyama, 2010).

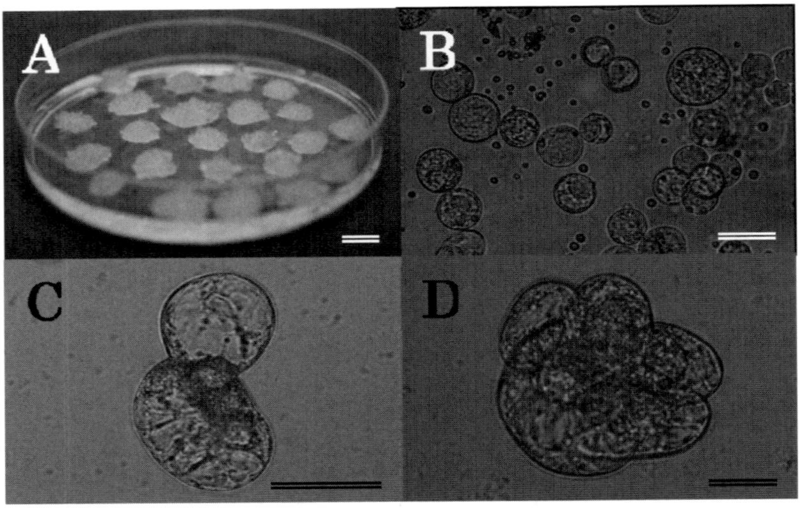

Figure 10. *Pinus luchuensis* protoplast isolation and culture. (A) Embryogenic tissue used for protoplast isolation. (B) Freshly isolated protoplasts. (C) A dividing protoplast after 4 days of culture. (D) Colony formation after 7 days of culture. *Bars*: 1 cm (A), 50 μm (B-D).

CONCLUSION

Of the 192 megagametophytes containing zygotic embryos cultured, only one resulted in a stable embryogenic cell line (0.5%). This low initiation rate was similar to the results reported for other pine species such as *Pinus banksiana* (0.37%) (Park et al. 1999), *Pinus patula* (0.04%) (Jones and van Staden, 1999), and *Pinus rigida* × *P. taeda* (0.44%) (Kim and Moon, 2007). The low initiation frequency of somatic embryogenesis is one of the key problems to resolve for practical applications. Although the condition of the mother tree and the quality of seeds have been reported to affect the frequency of somatic embryogenesis in several species, it is important in establishing productive embryogenic cultures, since seed genotypes and culture procedures may impact on the initiation frequencies of somatic embryogenesis (Park et al. 2006, Igasaki et al. 2003, Miguel et al. 2004, Klimaszewska et al. 2001, Lelu et al. 1999, Garin et al. 1998).

Somatic embryo maturation was obtained by culturing embryogenic tissue on media supplemented with maltose, PEG, ABA, and AC. The supplement of PEG to the media was essential to obtain high maturation frequencies of somatic embryos as reported for other Japanese pines (Maruyama et al. 2005a, 2005b, 2005d, 2007a, Hosoi et al. 2015, Maruyama and Hosoi, 2016a, 2016b). The stimulating effect of PEG on somatic embryo maturation in conifers has been extensively reported (Gupta and Pullman, 1991, Jain et al. 1995, Li et al. 1997, 1998, Ishii et al. 2001, Pullman et al. 2003, Jain and Gupta, 2005, Maruyama et al. 2002, 2005c, 2007b, Hosoi and Ishii, 2001, Maruyama and Hosoi, 2012a, Hosoi and Maruyama, 2016), and may be related to water stress induction similar to that generated by desiccation and to an increase in the accumulation of storage reserves (Roberts et al. 1990a, Attree et al. 1992, Misra et al. 1993). In the present study, the best result obtained on somatic embryo production (282 embryos per plate) represented an improvement of more than 25-fold compared to that achieved on a PEG-free medium (11 embryos per plate) (Figure 2).

Slow desiccation at high relative humidity was found to be effective in promoting the germination of somatic embryos. Desiccated embryos germinated rapidly and after 12 weeks of culture, more than 80% were converted into plants. In contrast, when somatic embryos without post-maturation treatment were placed directly on the germination medium, the root emergence of embryos occurred at a very low frequency (data not shown). Post-maturation treatment after the PEG-mediated maturation of somatic embryos has also been reported to successfully improve germination frequencies in *Picea engelmanni-glauca* (Roberts et al. 1990b), *Picea sitchensis* (Roberts et al. 1991), *Picea glauca* (Kong and Yeung, 1992; Kong and Yeung, 1995; Attree et al. 1995), *Larix* × *leptoeuropaea* (Lelu et al. 1995; Dronne et al. 1997), *Pinus patula* (Jones and van Staden, 2001), and *P. thunbergii* (Maruyama and Hosoi, 2012b). The regenerated somatic embryo-derived plants were successfully acclimatized and grew without any abnormal appearance in morphologic features.

We believe that the regeneration system described in this chapter represents a promising tool for efficient propagation and for future genetic engineering approaches to develop transgenic Ryukyumatsu resistance to the pinewood nematode.

ACKNOWLEDGMENT

The authors would like to acknowledge the Iriomote Tropical Forest Tree Breeding Technical Garden for their generous supply of seeds.

REFERENCES

Akiba, M. & Nakamura, K. (2005). Susceptibility of adult trees of the endangered species *Pinus armandii* var *amamiana* to pine wilt disease in the field. *Journal of Forestry Research, 10*, 3-7.

Attree, S. M., Dunstan, D. I. & Fowke, L. C. (1989). Initiation of embryogenic callus and suspension cultures, and improved embryo regeneration from protoplasts, of white spruce (*Picea glauca*). *Canadian Journal of Botany, 67*, 1790-1795.

Attree, S. M., Pomeroy, M. K. & Fowke, L. C. (1992). Manipulation of conditions for the culture of somatic embryos of white spruce for improved triacylglycerol biosynthesis and desiccation tolerance. *Planta, 187*, 395-404.

Attree, S. M., Pomeroy, M. K. & Fowke, L. C. (1995). Development of white spruce (*Picea glauca* (Moench.) Voss) somatic embryos during culture with abscisic acid and osmoticum, and their tolerance to drying and frozen storage. *Journal of Experimental Botany, 46*, 433-439.

Bonga, J. M. (2016). Conifer clonal propagation in tree improvement programs. In Y. S. Park, J. M. Bonga and H. K. Moon (Eds.), *Vegetative propagation of forest trees* (pp. 3–31). *NIFOS*: Seoul, Korea.

David, A., line, E. & David, H. (1995). Somatic embryogenesis in *Pinus caribaea*. In S. M. Jain, P. K. Gupta, R. J. Newton (Eds.), *Somatic embryogenesis in woody plants*, vol. *3*. (pp. 145-181). Kluwer Academic Publishers: Dordrecht.

Dronne, S., Label, P. & Lelu, M. A. (1997). Desiccation decreases abscisic acid content in hybrid larch (*Larix* × *leptoeuropaea*) somatic embryos. *Physiologia Plantarum*, *99*, 433-438.

Garin, E., Isabel, N. & Plourde, A. (1998). Screening of large numbers of seed families of *Pinus strobus* L. for somatic embryogenesis from immature and mature zygotic embryos. *Plant Cell Reports*, *18*, 37-43.

Gupta, P. K. & Pullman, G. S. (1991). *Method for reproducing coniferous plants by somatic embryogenesis using abscisic acid and osmotical potential variation.* U.S. Patent No. 5,036,007.

Hosoi, Y. & Ishii, K. (2001). Somatic embryogenesis and plantlet regeneration in *Pinus armandii* var. amamiana. In N. Morohoshi, and A. Komamine (Eds.), *Molecular Breeding of Woody Plants* (pp. 313–318). Elsevier Science: Amsterdam.

Hosoi, Y. & Maruyama, T. E. (2010). Protoplast culture from embryogenic tissue of Japanese black pine. *Proc Jap For Soc*, *121*, D10 (in Japanese).

Hosoi, Y., Kuramoto, N. & Maruyama, E. T. (2015). Screening RAPD primers to assess clonal fidelity in somatic embryos of Sawara cypress (*Chamaecyparis pisifera* Sieb. *et* Zucc.) and field performance of somatic embryo-derived trees. *Plant Biotechnology*, *32*, 149-155.

Hosoi, Y. & Maruyama, T. E. (2012). Plant regeneration from embryogenic tissue of *Pinus luchuensis* Mayr, an endemic species in Ryukyu Island, Japan. *Plant Biotechnology*, *29*, 401-406.

Hosoi, Y. & Maruyama, T. E. (2016). Somatic Embryogenesis in Sawara cypress (*Chamaecyparis pisifera* Sieb. *et* Zucc.). In A. Mujib (Eds), *Somatic embryogenesis in ornamentals and its applications* (pp. 41–53). Springer: India.

Igasaki, T., Sato, T., Akashi, N., Mohri, T., Maruyama, E., Kinoshita, I., Walter, C. & Shinohara, K. (2003). Somatic embryogenesis and plant regeneration from immature zygotic embryos of *Cryptomeria japonica* D. Don. *Plant Cell Reports*, *22*, 239–243.

Ishii, K., Maruyama, E. & Hosoi, Y. (2001). Somatic embryogenesis of Japanese conifers. In N. Morohoshi, and A. Komamine (Eds.), *Molecular breeding of woody plants* (pp. 297–304). Elsevier Science: Amsterdam.

Jain, S. M., Gupta, P. K. & Newton, R. J. (1995). Somatic embryogenesis. In *woody plants*, vol. *3*, *Gymnosperms*. Kluwer Academic Publishers: Dordrecht.

Jain, S. M. & Gupta, P. K. (2005). *Protocol for somatic embryogenesis in woody plants*. Springer: Dordrecht.

Jones, N. B. & van Staden, J. (1999). Somatic embryogenesis in *Pinus patula* Scheide et Deppe. In S. M. Jain, P. K Gupta, and R. J. Newton (Eds.), *Somatic embryogenesis in woody plants*, vol. *4*. (pp. 431-447). Kluwer Academic Publishers: Dordrecht.

Jones, N. B. & van Staden, J. (2001). Improved somatic embryo production from embryogenic tissue of *Pinus patula*. *In vitro Cellular & Developmental Biology- Plant*, *37*, 543-549.

Kim, Y. W. & Moon, H. K. (2007). Regeneration of plant by somatic embryogenesis in *Pinus rigida* × *P. taeda*. *In vitro Cellular & Developmental Biology- Plant*, *43*, 335-342.

Klimaszewska, K., Park, J. S., Overton, C., Maceacheron, I. & Bonga, J. M. (2001). Optimized somatic embryogenesis in *Pinus strobus* L. In *In vitro Cellular & Developmental Biology- Plant*, *37*, 392–399.

Kong, L. & Yeung, E. (1992). Development of white spruce somatic embryos: II. Continual shoot meristem development during germination. *In vitro Cellular & Developmental Biology- Plant*, *28*, 125-131.

Kong, L. & Yeung, E. (1995). Effects of silver nitrate and polyethylene glycol on white spruce (*Picea glauca*) somatic embryo development: enhancing cotyledonary embryo formation and endogenous ABA content. *Physiologia Plantarum*, *93*, 298-304.

Lelu, M. A., Klimaszewska, K., Pflaum, G. & Bastien, C. (1995). Effect of maturation duration on desiccation tolerance in hybrid larch (*Larix* ×

leptoeuropaea Dengler) somatic embryos. *In vitro Cellular & Developmental Biology- Plant, 31*, 15-20.

Lelu, M. A., Bastien, C., Drugeault, A., Gouez, M. L. & Klimaszewska, K. (1999). Somatic embryogenesis and plantlet development in *Pinus sylvestris* and *Pinus pinaster* on medium with and without growth regulators. *Physiologia Plantarum, 105*, 719–728.

Li, X. Y., Huang, F. H. & Gbur, E. E. Jr. (1997). Polyethylene glycol-promoted development of somatic embryo in loblolly pine (*Pinus taeda* L.). *In vitro Cellular & Developmental Biology- Plant, 33*, 184-189.

Li, X. Y., Huang, F. H., Murphy, J. B. & Gbur, E. E. Jr. (1998). Polyethylene glycol and maltose enhance somatic embryo maturation in loblolly pine (*Pinus taeda* L.). *In vitro Cellular & Developmental Biology- Plant, 34*, 22-26.

Maruyama, T. E. & Hosoi, Y. (2012a). Somatic embryogenesis and efficient plant regeneration in Japanese cypresses. In K. Sato (Eds.) *Embryogenesis* (pp. 387–402). InTech, Rijeka: Croatia.

Maruyama, T. E. & Hosoi, Y. (2012b). Post-maturation treatments improves and synchronizes somatic embryo germination of three species of Japanese pines. *Plant Cell, Tissue and Organ Culture, 10*, 45-52.

Maruyama, T. E. & Hosoi, Y. (2014). Plant production in Japanese pines via somatic embryogenesis. In K. G. Ramawat, J. M. Merillon, and M. R. Ahuja (Eds.), *Tree biotechnology* (pp. 251–261). CRC Press: Florida.

Maruyama, T. E. & Hosoi, Y. (2016a). Somatic embryogenesis in Japanese black pine (*Pinus thunbergii* Parl.). In A. Mujib (Eds.), *Somatic embryogenesis in ornamentals and its applications* (pp. 27–39). Springer: India.

Maruyama, T. E. & Hosoi, Y. (2016b). Somatic embryogenesis and plant propagation in Japanese black pine (*Pinus thunbergii* Parl.) and Japanese red pine (*Pinus densiflora* Zieb. *et* Zucc.). In Y. S. Park, J. M. Bonga, and H. K. Moon (Eds.), *Vegetative propagation of forest trees* (pp. 623–638). NIFOS: Seoul, Korea.

Maruyama, E., Tanaka, T., Hosoi, Y., Ishii, K. & Morohoshi, N. (2000). Embryogenic cell culture, protoplast regeneration, cryopreservation, biolistic gene transfer and plant regeneration in Japanese cedar (*Cryptomeria japonica* D. Don). *Plant Biotechnology*, *17*, 281-296.

Maruyama, E., Hosoi, Y. & Ishii, K. (2002). Somatic embryogenesis in *Sawara cypress* (*Chamaecyparis pisifera* Sieb. et Zucc.) for stable and efficient plant regeneration, propagation and protoplast culture. *Journal of Forest Research*, *7*, 23-34.

Maruyama, E., Hosoi, Y. & Ishii, K. (2005a). Somatic embryo production and plant regeneration of Japanese black pine (*Pinus thunbergii*). *Journal of Forest Research*, *10*, 403-407.

Maruyama, E., Hosoi, Y. & Ishii, K. (2005b). Propagation of Japanese red pine (*Pinus densiflora* Zieb. et Zucc.). *Propagation of Ornamental Plants*, *4*, 199-204.

Maruyama, E., Ishii, K. & Hosoi, Y. (2005c). Efficient plant regeneration of Hinoki cypress (*Chamaecyparis obtusa* Sieb. et Zucc.) via somatic embryogenesis. *Journal of Forest Research*, 10, 73-77.

Maruyama, E., Hosoi, Y. & Ishii, K. (2005d). Sawara cypress (*Chamaecyparis pisifera* Zieb. et Zucc.). In S. M. Jain, and P. K. Gupta (Eds.), *Protocol for somatic embryogenesis in woody plants* (pp. 459-468). Springer: Netherlands.

Maruyama, E., Hosoi, Y. & Ishii, K. (2007a). Somatic embryogenesis and plant regeration in yakutanegoyou, *Pinus armandii* Franch. var. *amamiana* (Koidz.) Hatusima, an endemic and endangered species in Japan. *In vitro Cellular & Developmental Biology - Plant*, *43*, 28-34.

Maruyama, E., Hosoi, Y. & Ishii, K. (2007b). Polyethylene glycol enhance somatic embryo production in Japanese cedar (*Cryptomeria japonica* D. Don). *Propagation of ornamental plants*, *2*, 57-61.

Miguel, C., Gonçalves, S., Tereso, S., Marum, L., Maroco, J. & Olivera, M. M. (2004). Somatic embryogenesis from 20 open-pollinated families of Portuguese plus trees of maritime pine. *Plant Cell, Tissue and Organ Culture*, *76*, 121-130.

Misra, S., Attree, S. M., Leal, I. & Fowke, L. C. (1993). Effects of abscisic acid, osmoticum and desiccation on synthesis of proteins during the development of white spruce somatic embryos. *Annals Botany*, *71*, 11-22.

Mota, M. M., Braasch, H., Bravo, M. A., Penas, A. C., Burgermeister, W., Metge, K. & Sousa, E. (1999). First report of *Bursaphelenchus xylophilus* in Portugal and in Europe. *Nematology*, *1*, 727–734.

Nagao, A. (1983). Differences of flower initiation of *Cryptomeria japonica* under various alternating temperatures. *Journal of the Japanese Forest Society*, *65*, 335-338 (in Japanese).

Park, Y. S., Barrett, J. D. & Bonga, J. M. (1998). Application of somatic embryogenesis in high-value clonal forestry: deployment, genetic control, and stability of cryopreserved clones. *In vitro Cellular & Developmental Biology- Plant*, *34*, 231-239.

Park, Y. S., Bonga, J. M., Cameron, S. I., Barrett, J. D., Forbes, K., DeVerno, L. L. & Klimaszewska, K. (1999). Somatic embryogenesis in jack pine (*Pinus banksiana* Lamb). In S. M. Jain, P. K. Gupta, and R. J. Newton (Eds.), *Somatic embryogenesis in woody plants*, vol. *4*. (pp. 491-504). Kluwer Academic Publishers: Dordrecht.

Park, Y. S., Lelu-Walter, M. A., Harvengt, L., Trontin, J. F., MacEacheron, I., Klimaszewska, K. & Bonga, J. M. (2006). Initiation of somatic embryogenesis in *Pinus banksiana*, *P. strobus*, *P. pinaster*, and *P. sylvestris* at three laboratories in Canada and France. *Plant Cell, Tissue and Organ Culture*, *86*, 87-101.

Pullman, G. S., Johnson, S., Peter, S., Cairney, J. & Xu, N. (2003). Improving loblolly pine somatic embryo maturation: comparison of somatic and zygotic embryo morphology, germination, and gene expression. *Plant Cell Reports*, *21*, 747-758.

Roberts, D. R., Flinn, B. S., Webb, D. T., Webster, F. B. & Sutton, B. C. S. (1990a). Abscisic acid and indole-3-butyric acid regulation of maturation and accumulation of storage proteins in somatic embryos of interior spruce. *Physiologia Plantarum*, *78*, 355-360.

Roberts, D. R., Sutton, B. C. S. & Flinn, B. S. (1990b). Synchronous and high frequency germination of interior spruce somatic embryos following partial drying at high relative humidity. *Canadian Journal of Botany*, *68*, 1086-1090.

Roberts, D. R., Lazaroff, W. R. & Webster, F. B. (1991). Interaction between maturation and high relative humidity treatments and their effects on germination of Sitka spruce somatic embryos. *Journal of Plant Physiology*, *138*, 1-6.

Park, Y. S., Barrett, J. D. & Bonga, J. M. (1998). Application of somatic embryogenesis in high-value clonal forestry: deployment, genetic control, and stability of cryopreserved clones. *In vitro Cellular & Developmental Biology - Plant*, *34*, 231-239.

Park, Y. S., Bonga, J. M., Cameron, S. I., Barrett, J. D., Forbes, K., DeVerno, L. L. & Klimaszewska, K. (1999). Somatic embryogenesis in jack pine (*Pinus banksiana* Lamb) (pp. 491-504). In S. M. Jain, P. K. Gupta, and R. J. Newton (Eds.), *Somatic embryogenesis in woody plants*, vol. 4. (pp. 491-504). Kluwer Academic Publishers: Dordrecht.

Tautorus, T. E., Attree, S. M., Fowke, L. C. & Dunstan, D. I. (1990). Somatic embryogenesis from immature and mature zygotic embryos, and embryo regeneration from protoplasts in black spruce (*Picea mariana* Mill.). *Plant Science*, *67*, 115-124.

Togashi, K. & Shigesada, N. (2006). Spread of the pinewood nematode vectored by the Japanese pine sawyer: modeling and analytical approaches. *Population Ecology*, *48*, 271–283.

PART VI: *IN VITRO* PROPAGATION OF ORNAMENTAL MONOCOTS

In: Micropropagation of Ornamental Plants ISBN: 978-1-53614-541-0
Editor: Taoufik Bettaieb

Chapter 8

SOMATIC EMBRYOGENESIS AS A TOOL FOR PROPAGATION OF PALM TREES (ARECACEAE): PRINCIPLES, STRATEGIES AND CHALLENGES

Inaê Mariê de Araújo Silva*[1], *PhD
and Jonny Everson Scherwinski-Pereira*[2,*], *PhD
[1]Department of Forest Engineering,
University of Brasília, Brasília, Brazil
[2]Embrapa Genetic Resources & Biotechnology, Brasília, Brazil

ABSTRACT

For palm trees, the most commonly used, widely studied and efficient technique for *in vitro* propagation of elite material is somatic embryogenesis, which has already been applied successfully in dozens of species and varieties. This technique reflects the potential for unique somatic cell development of plants, which – through appropriate *in vitro* conditions – results in the transition of differentiated somatic cells into

* Corresponding Author Email: jonny.pereira@embrapa.br.

bipolar structures similar to zygotic embryos that regenerate plants genetically identical to the source material. One of the most striking characteristics of somatic embryogenesis is the variety of physical and chemical factors that induce the process, with an emphasis on growth regulators, most notably auxins and cytokinins, and/or various stress treatments. In addition to the inductive factors, a broad group of different factors influences the process, especially the genotype, the donor plant's physiology, the type and physiological state of explant development, and composition of the culture medium. Aside from being multifactorial, somatic embryogenesis in ornamental palms involves a set of phases (induction, differentiation, maturation, germination/conversion and acclimatization) detailed in this review. Examples of successful use of the technique in multiplication of ornamental palms, as well as the bottlenecks inherent in the technique, are also explained here. Despite the exhaustive literature on the subject, somatic embryogenesis is still little elucidated, especially regarding the understanding of the processes involved in somatic-embryogenic cell transition and in the final stages of the process (plant maturation/regeneration).

Keywords: propagation, plant growth regulators, morphogenesis, micropropagation, somatic embryos

INTRODUCTION

The Arecaceae family, generally known as the palm tree family, is one of the main evolutionary trunks of monocots, being composed of 252 genera and around 2,600 species distributed worldwide (Dransfield et al. 2008). Representing the third most important botanical family for humans, palm trees stand out in their productivity and multiplicity of uses (Johnson, 1998), especially the ornamental use in a wide range of climatic conditions (tropical, Mediterranean, and subtropical climates) around the world. The emblematic use of different palm trees composing the most diverse landscapes, such as *Cocos nucifera* in the United States (Hawaii and Southern Florida) (Broschat et al. 2014), *Phoenix dactylifera* in southern Europe and the United States (Arizona, California, Nevada, and Florida), (Chao and Krueger, 2007) and *Roystonea oleraceae* in Brazil (Araújo and

Silva, 2010), exemplifies the ornamental use of palm trees around the globe.

The commercial use of palm trees for the different uses, including ornamental, requires the development of technologies that include the selection of superior genotypes and propagation thereof by efficient systems. However, the propagation of many palm trees through the sexual pathway may be inefficient, due to low percentages of germination and reduced viability of seeds (Broschat et al. 2014; Ree and Guerra, 2015), plus the resulting genetic heterogeneity, which is an unfavorable factor in terms of commercial-scale production (Saldanha and Martins-Corder, 2012).

Figure 1. List of palm species with published somatic embryogenesis protocols [adapted from Ree and Guerra] (2015)] and ornamental palms: A) *Elaeis guineensis*, B) *Phoenix* sp., C) *Cocos nucifera*, D) *Acrocomia aculeata*, E) *Attalea speciosa*, and F) *Syagrus oleracea*. (Source A-D, F: IMA Silva-Cardoso, E: EOL Saleh).

It is also noteworthy that palm trees present singular growth habits, without secondary growth (Tomlinson and Huggett, 2012), and tiller production limited to a few species (Nascimento and Oliveira, 2011), thereby limiting the application of traditional methods of vegetative propagation and the consequent use of the superior genotypes (Moura et al. 2009; Saldanha and Martins-Corder, 2012). In this context, somatic embryogenesis, an in-vitro culture technique, is projected as a promising alternative for the clonal propagation of palm trees (Saldanha and Martins-Corder, 2012).

Somatic embryogenesis, first reported in carrots (*Daucus carota* L.) (Reinert, 1958; Steward et al. 1958) and described for the first time in palm trees by Staritsky (1970) for the species *Elaeis guineensis* (African oil palm), is the tissue culture technique most widely used in the vegetative propagation of palms (Steinmacher et al. 2007). According to Gabr and Tisserat (1985), the mass cloning of palms is only possible thanks to the technique of somatic embryogenesis. Since its first description in palm trees, there have been reports of the development of somatic embryogenesis protocols for 18 species and dozens of varieties (Ree and Guerra, 2015) (Figure 1). Recently, two more species were added to this group, *Attalea speciosa* (Saleh, 2016) and *Syagrus oleracea* (Cardoso-Silva, 2018).

This technique, considered an example of the high plasticity exhibited by plant cells (Fehér, 2015), has been described as a complex multi-step and multi-factorial process involving the transition from somatic cells to structures similar to zygotic embryos (Emons, 1994), with no fusion of gametes (Williams and Maheswaran, 1986). These structures are bipolar and have no vascular connection with the tissue of origin (Von Arnold et al. 2002), and are able to regenerate whole plants genetically identical to the plant supplying the initial material. In general, this transition process involves the dedifferentiation of cells, activation of the cell division cycle and reorganization of physiology, metabolism, and patterns of gene expression (Fehér et al. 2003; Yang and Zahng, 2010).

INDUCTIVE FACTORS OF SOMATIC EMBRYOGENESIS: GROWTH REGULATORS AND STRESSES

One of the most striking features of somatic embryogenesis is the variety of physical and chemical factors that induce the process, which include growth regulators, mainly auxins and cytokinins, and/or various stress treatments (Fehér, 2006; Karami and Said, 2010; Zavattieri et al. 2010; Fehér, 2015; Nic-Can et al. 2016), such as osmotic (Karami et al. 2006; You et al. 2006; Tao et al. 2012; Moon et al. 2015; Muilu-Mäkelä et al. 2015), saline (Ibraheem et al. 2012; Taha and Hassan, 2014; Al-Khayri and Ibraheem, 2014), hydric (Al-Khayri and Ibraheem, 2014; Pereira et al. 2016), temperature (Kamada et al. 1994; Kikuchi et al. 2006; Moon et al. 2015; Pereira et al. 2016), heavy metals (Patnaik et al. 2005), injuries (Santarem et al. 1997; Cheong et al. 2002; Mozgová et al. 2017), nutritional (Mihaljevic et al. 2011; Moon et al. 2015; Krishnan and Siril, 2017), and oxidative (Zhou et al. 2016), among others Gaj (2004) points out – in addition to the proper choice of the explant, which should be a source of competent cells – the determination of the inductive factors as prerequisites for the successful establishment of a somatic embryogenesis system.

Growth regulators and various stressors play a central role in mediating the signal transduction cascades that lead to gene reprogramming (Dudits et al. 1991). This fact results in a series of cell divisions that induce polarized growth or disorganized growth of calli, leading to direct and indirect somatic embryogenesis, respectively (Sharp et al. 1980; Dudits et al. 1991).

It is known that auxins and cytokinins are considered central regulators of the processes of cell division as well as dedifferentiation and redifferentiation, involved in the acquisition of embryogenic competence (Zeng et al. 2007). Among these regulators, auxins stand out as the plant growth regulator that is most commonly used in the induction of somatic embryogenesis, as well as in callus induction and control of organ regeneration (Nic-Can and Loyola-Vargas, 2016).

In several plant families, not unlike Arecaceae, auxin analogues have played a crucial role in the embryogenic process (Viñas and Jiménez, 2011), especially 2,4-dichlorophenoxyacetic acid (2,4-D) (Fehér et al. 2003; Karami and Said, 2010; Nic-Can and Loyola-Vargas, 2016), usually used in high concentrations in the induction phase in different palm trees, such as *Cocos nucifera* (Karunaratne and Periyapperuma, 1989; Sané et al. 2006; Nguyen et al. 2015; Sandoval-Cancino et al. 2016), *E. guineensis* (Teixeira et al. 1993; Guedes et al. 2011), and *Phoenix dactylifera* (Tisserat, 1979; Veramendi and Navarro, 1996; Othmani et al. 2009; Sané et al. 2012; Kurup et al. 2014; Mazri et al. 2017). Saleh and Scherwinski-Pereira et al. (2016) point out that out of a total of 20 protocols of somatic embryogenesis in palm trees analyzed based on articles published between 2000 and 2015, 70% of them used 2,4-D auxin in embryogenic induction.

The high embryogenic efficacy of auxins, specifically 2,4-D, may be a consequence of the dual role played by this regulator, since it behaves like an auxin, acting directly upon or modifying the intracellular metabolism of endogenous auxins (Fehér et al. 2003; Kikuchi et al. 2006); on the other hand, 2,4-D is also considered a stressor (Bhat et al. 2015; Nic-Can and Loyola-Vargas, 2016). According to Karami and Said (2010), when exogenous auxins are applied, they behave more as stressors than regulators, per se, and this issue is supported by the shared expression of genes in response to auxins and to different abiotic stresses.

The relationship between different stressors and somatic embryogenesis has been widely explored in recent years, mainly after the discovery of shared genes between somatic embryogenesis and stress response signaling pathways (Davletova et al. 2001; Gliwicka et al. 2013; Jin et al. 2014; Leljak-Levanić et al. 2015; Wickramasuriya and Dunwell, 2015; Cheng et al. 2016; Fraga et al. 2016a; Nowak and Gaj, 2016); this suggests that somatic embryogenesis is an adaptive response by the plant genome to the stress of *in vitro* culture (Dudits et al. 1995; Karami and Said, 2010; Jin et al. 2014; De-La-Peña et al. 2015; Nowak and Gaj, 2016). According to Fehér (2015), although this view of the role of stress on somatic embryogenesis is currently widely accepted, the underlying mechanisms have not yet been fully elucidated. According to Altamura et al. (2016), when the degree of stress exceeds cell tolerance, the cells collapse and die; on the other hand, when the level of stress is lower, there is an increase in the metabolic activity and induction of the adaptation process, including the reprogramming of the genetic expression and cell reorganization.

The use of auxins other than 2,4-D has also been reported in palms, with different efficiency results, such as 3,6-dichloro-2-methoxybenzoic acid (Dicamba) (Chehmalee and Te-Chato, 2008; Wang et al. 2010; Granja et al. 2018) and 4-amino-3,5,6-trichloropicolinic acid (Picloram) (Karun et al. 2004; Steinmacher et al. 2007; Thuzar et al. 2011; Pádua et al. 2013; Luis and Scherwinski-Pereira, 2014; Padilha et al. 2015; Bonetti et al. 2016; Mazri et al. 2017; Granja et al. 2018); as well as the combination between auxins 2,4-D + Picloram (Yusnita and Hapsoro, 2011; Jayanthi et al. 2015). Picloram, for example, has provided good results in the induction of somatic embryogenesis in *Bactris gasipaes* (Valverde et al. 1987; Steinmacher et al. 2007; Steinmacher et al. 2007a, b), *Calamus merrillii* and *C. subinermis* (Goh et al. 1999; Goh et al. 2001), and *Euterpe oleracea* (Scherwinski-Pereira et al. 2012). On the other hand, Dicamba was effective in callus induction in *Areca catechu* (Wang et al. 2003).

Also noteworthy are reports of the use of cytokinins such as 6-benzylaminopurine (BAP), 2-isopentenladenine (2-iP) and Thidiazuron (TDZ), often at low concentrations, whether or not combined with auxins, in the induction of somatic embryogenesis in palm trees (Gallo-Meagher and Green, 2002; Eshraghi et al. 2005; Almeida and Almeida, 2006; Perera et al. 2009; Boufis et al. 2014; Ali and Ali, 2015; Zayed and Elbar, 2015; Mazri et al. 2017; Granja et al. 2018). According to Bhatia and Bera (2015), among the cytokinins, TDZ stands out as one of the most commonly used in somatic embryogenesis.

Another class of growth regulators, polyamines, has also provided good results in somatic embryogenesis in palms such as *C. nucifera* (Adkins et al. 1998; Rajesh et al. 2014), *P. dactylifera* (Hegazy and Aboshama, 2010) and *E. guineensis* (Rajesh et al. 2003; Corrêa et al. 2016). Several authors also report positive effects of the exogenous application of polyamines in the various phases of somatic embryogenesis in species from other botanical families, such as *Panax ginseng* (Kevers et al. 2000), *Daucus carota* (Takeda et al. 2002), *Momordica charantia* (Paul et al. 2009), *Brassica napus* (Ahmadi et al. 2014), *Gossypium hirsutum* (Cheng et al. 2015), *Saccharum* sp. (Reis et al. 2016) and *Oryza sativa* (Tan et al. 2017). On the other hand, there are reports that exogenous application of polyamines does not promote effects on somatic embryogenesis, as observed in *Picea abies* (Vondráková et al. 2015), or causes inhibitory effects at some specific stage of the process, as mentioned regarding *Mangifera indica* (Litz and Shaffer, 1987) and *Solanum melongena* (Fobert and Webb, 1988).

The physiological effects of polyamines on somatic embryogenesis still remain little elucidated (Cheng et al. 2015; Reis et al. 2016); they are commonly related to cell division (Paul et al. 2009) and rapid growth (Kackar and Shekhawat, 2007), cell elongation and development of meristematic centers (Cvikrová et al. 1998), and protection against the stressful *in vitro* environment (Reis et al. 2016).

This versatility of inductive factors can be partly explained by the interaction of the inductive condition with the endogenous levels of hormones in the explants (Jiménez, 2001, 2005), particularly the hormones indole-3-acetic acid (AIA) and abscisic acid (ABA), considered crucial in determining the embryogenic potential of the explants (Gaj, 2004).

The endogenous levels of gibberellin (GA), on the other hand, exhibit a negative relation with somatic embryogenesis, according to Wang et al. (2013), Altamura et al. (2016), and Zayed et al. (2016). In addition to these hormones, ethylene – commonly synthesized in response to various biotic and abiotic stresses (Wang et al. 2002)– has also been highlighted, although its role in somatic embryogenesis has not yet been fully elucidated, due to the ambiguous result (as both inhibitor and promoter) reported in different species and cultivation systems (Altamura et al. 2016). Bai et al. (2013) suggest that the effects of ethylene on somatic embryo formation depend on the genetic origin of the explants and the culture conditions. According to Loyola-Vargas and Ochoa-Alejo (2016), these two latter hormones play more relevant roles in the final stages of embryogenesis.

Ooi et al. (2012), in a study involving *E. guineensis* aimed at correlating the potential for somatic embryogenesis of leaf explants with alterations in the expression of genes responsive to hormones, suggested that endogenous auxin levels may be higher in highly embryogenic palms. In contrast, Zhou et al. (2017), studying *Cunninghamia lanceolata*, reported an inverse relationship between endogenous auxin levels and embryogenic competence. Although there are several reports of the increase of endogenous hormones in embryogenic systems, these internal hormones are not the only factors involved in determining response/stimulus inductive specificity and, apparently, are not adequate markers of embryogenic potential in explants of which *in vitro* behavior is unknown (Jiménez, 2001).

ROUTES OF DEVELOPMENT AND UNICELLULAR/MULTICELLULAR ORIGINS OF SOMATIC EMBRYOS

The formation of somatic embryos can be obtained through two development routes: direct or indirect. In direct somatic embryogenesis, the cells already have embryogenic competence (specific pre-embryogenic cells) and require a minimal genetic reprogramming for embryo formation, with no dedifferentiation stage (Sharp et al. 1980; Williams and Maheswaran, 1986; Yeung, 1995). This is certainly because they present totipotency-related genes accessible to transcription (Horstman et al. 2017), differently from what occurs in indirect somatic embryogenesis. In this route of embryogenesis, the cells require a greater genetic reprogramming to acquire the state of embryogenic competence (induction of certain embryogenic cells), which is preceded by the proliferation of callus cells (Sharp et al. 1980; Williams and Maheswaran, 1986; Yeung, 1995). According to Sharp et al. (1980) and Williams and Maheswaran (1986), in indirect somatic embryogenesis, growth regulators play a crucial role for re-entering the cell division process, as well as for determining the embryogenic state.

The main factors involved in each route, in turn, depend on the source and the physiological state of the explant used (Williams and Maheswaran, 1986), and on the type and concentration of plant growth regulators added to the system (Quiroz-Figueroa et al. 2006; Tagipur et al. 2016). According to Horstman et al. (2017), it appears that the definition of the occurrence of direct or indirect somatic embryogenesis is more related to the combination of the explant development context with the culture environment, rather than the development distance of the embryonic state.

This model of embryogenesis, with previous formation of cellular aggregates or calli (disorganized cell masses, with different degrees of differentiation) (Ikeuchi et al. 2013; Perianez-Rodriguez et al. 2014) inserts the concept of cell redifferentiation into the context, characterized as the regeneration of plants starting from calli, originating from the dedifferentiation of somatic cells or originating from pre-existing meristematic cells (Bhojwani and Dantu, 2013).

During indirect somatic embryogenesis, not only the embryogenic calli, capable of forming somatic embryos, but also and non-embryogenic calli (Figure 2) are formed, and are distinguishable by aspects related especially to morphology and coloring (Na et al. 2007; Von Arnold et al. 2002; Quiroz-Figueroa et al. 2006; Yang and Zahng, 2010; Borji et al. 2017), which also vary among species. Shang et al. (2009) also point out that obvious differences in the pattern of cell arrangement on the surface of the calli, in the histological characteristics, as well as in the cytological ultrastructure can potentially be investigated in a comparative way between cells of embryogenic calli and non-embryogenic calli.

Embryogenic callus cells usually exhibit specific characteristics typical of meristematic cells (Halperin and Jensen, 1967; Talapatra et al. 2016), such as an isodiametric shape, dense cytoplasmic content (Purohit, 2013), reduced size, large nuclei with prominent nucleoli (Namasivayam, 2007), small/fragmented vacuoles (Halperin and Jensen, 1967; Yang and Zahng, 2010), and accumulation of starch (Williams and Maheswaran, 1986; Yeung, 1995). Additionally, they are metabolically active cells, with intense mitogenic activity (Yeung 1995; Petrek et al. 2015).

However, some authors have proposed that meristematic cells are distinguished from embryogenic cells in ultrastructural characteristics related to nuclei, in such a way that the meristematic cells have spherical nuclei and many nucleoli, while embryogenic cells exhibit irregularly shaped nuclei and a single prominent nucleolus (Verdeil et al. 2007). Also according to these authors, embryogenic cells exhibit little heterochromatin and high euchromatin, exactly the opposite of what was observed in meristematic cells. Despite the foregoing differences, the two terms have been used as synonyms in a significant part of the literature.

Non-embryogenic callus cells are usually vacuolated (Ronchi and Giorgetii, 1995; Samaj et al. 1999; Kurczynska et al. 2012; Borji et al. 2017), elongated (Pasternak et al. 2002), dispersed, with a low nucleus/cytoplasm ratio, (Silveira et al. 2013) and few organelles (Na et al. 2007), which – according to Carvalho et al. (2013) – can be interpreted as indicative of low metabolic activity.

Moreover, biochemical analyses of concentrations of endogenous hormone, such as AIA, ABA and polyamines, phenolic components and amino acids can be used to better distinguish between embryogenic callus and non-embryogenic callus strains, as performed by Nascimento-Gavioli et al. (2017) in *B. gasipaes*.

In palm trees, most studies mention the occurrence of indirect somatic embryogenesis (Viñas and Jiménez, 2011), as verified in *C. nucifera* (Saénz et al. 2006), *Acrocomia aculeata*, (Moura et al. 2009) and *E. guineensis* (Balzon et al. 2013). Somatic embryos originating from the indirect pathway show a high probability of multicellular origin (Gaj, 2004), a fact observed in the aforementioned studies. The origin of unicellular somatic embryos has also been reported in palm trees (Kanchanapoom and Domyoas, 1999; Luis and Scherwinski-Pereira, 2014), as well as the two pathways in the same regeneration system, as reported in *B. gasipaes* (Almeida et al. 2012), *E. oleracea* (Scherwinski-Pereira et al. 2012), and *P. dactylifera* (Sané et al. 2006).

According to Almeida et al. (2012), embryos originated via the unicellular route (asymmetric division of a single cell) (Figure 2E) are consequent of the early detachment of the tissues of origin as due to the rupture of the epidermis, and are formed from cells external to the tissues, presenting easy detachment of the parent tissue and, in some cases, exhibit suspensors. On the other hand, those originating from the multicellular route are internally formed by a group of meristematic cells, which makes it difficult to isolate the explant (Figure 2F). According to Puigderrajols et al. (2001), the origin of the somatic embryo is a pertinent aspect with respect to the genetic uniformity of the resulting plants, such that the unicellular origin is most desired.

Although direct somatic embryogenesis is less frequent in palm trees, as in most plant families (Gaj, 2004), this embryogenic route has the additional advantages of reducing the time required for plant propagation (El Bar and El Dawayati, 2014) and decreasing the occurrence of genetic changes(variação somaclonal) (Sivanesan et al. 2011; Jayanthi et al. 2011; Ibrahim et al. 2013), as well as the absence of accumulation of phenolic compounds in the medium and generally low contamination rates(Muniswamy and Sreenath, 1996; Herrera et al. 2011).

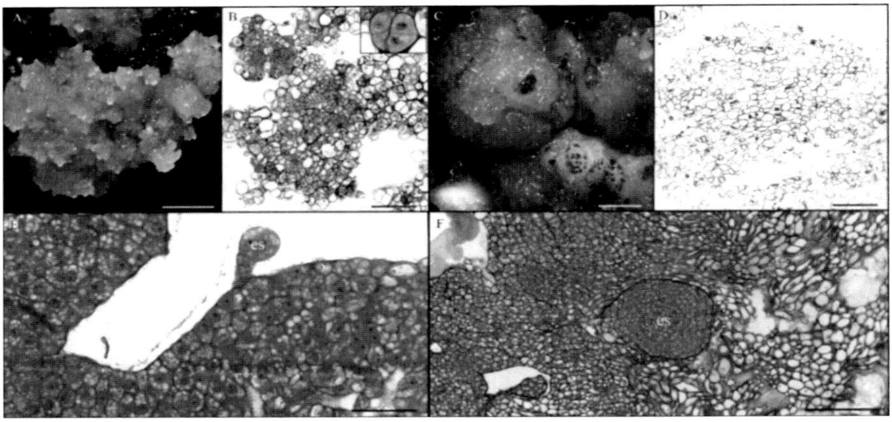

Figure 2. Morpho-anatomic aspects of somatic embryogenesis of gueroba palms [*Syagrus oleracea* Mart. (Becc.)]. Friable embryogenic callus B: Anatomy of friable embryogenic callus; note the proembryo (detalhe quadrado lateral). C: Non-embryogenic mucilaginous callus. D: Anatomy of non-embryogenic mucilaginous callus. E: Somatic embryo of unicellular origin. F: Somatic embryo of multicellular origin. Abbreviations: (es) somatic embryos. Bars = A, C: 2mm; D: 500 μm; B, F: 200 μm and E: 50 μm.

In palm trees, among the reports of obtaining somatic embryos without the intermediate calli phase, one can mention Jayanthi et al. (2011) starting from cotyledonary nodes of *E. guineensis*, and Sarasan et al. (2002) using haustoria, plumula and radicles of *Hyophorbe lagenicaulis* seedlings as explants. Direct somatic embryogenesis has also been reported in *P. dactylifera* by Sidky and

Eldawyati (2012) starting from immature inflorescences, and by El Bar and El Dawayati (2014) from young leaf explants. It is worth noting that despite these records, direct somatic embryogenesis in palm trees requires more detailed investigations.

A specific type of somatic embryogenesis – secondary somatic embryogenesis characterized by the production of somatic embryos from primary somatic embryos (Raemakers et al. 1995; Te-Chato and Hilae, 2007; Elmeer, 2013; Rocha et al. 2016) – has also been described in several species of palm: *A. catechu* (Wang et al. 2010), *B. gasipaes* (Steinmacher et al. 2011), *C. merrillii* e *C. subinermis* (Goh et al. 2001), *C. nucifera* (Buffard-Morel et al. 1992; Pérez-Núñez et al. 2006), *E. edulis* (Guerra and Handro, 1991), *E. oleracea* (Freitas et al. 2016), *E. guineensis* (Rajesh et al. 2003), and *P. dactylifera* (Othmani et al. 2009). According to Raemakers et al. (1995), this type of embryogenesis, compared to primary embryogenesis, has the advantages of improving the rates of multiplication, repeatability, independence of the explant source, and maintenance of the embryogenic capacity for relatively long time periods through repeated cycles of secondary embryogenesis. Furthermore, embryos obtained via secondary somatic embryogenesis may often exhibit better development than those obtained via primary somatic embryogenesis (Pérez-Núñez et al. 2006).

Stages of Somatic Embryogenesis

In general, protocols of somatic embryogenesis of palm trees involve the following phases: induction, differentiation, maturation, germination/ conversion and acclimatization. It noteworthy that not all protocols for somatic embryogenesis in palm trees differentiate between the stages, determining different treatments for each one of them. In fact, there is some confusion in the literature regarding the terms used to name them, largely because of the asynchronicity typical of the process of somatic embryogenesis.

Calli that contain embryogenic cells resulting from the induction phase, mediated by growth regulators and/or different stresses, usually proliferate and form proembriogenic masses (Halperin, 1966; Von Arnold et al. 2002; Yang and Zahng, 2010). These masses are usually characterized by the physical or physiological isolation of their immediate surroundings due to thickening of the cell wall (callose) (Dubois et al. 1990) and the absence or reduction of the number of plasmodesmata (Verdeil et al. 2001). According to Verdeil et al. (2001), this physical isolation can facilitate the reprogramming of the genomic and cell functions necessary for the acquisition of totipotency and embryogenic competence. It is worth mentioning that cell isolation is not a rule in in-vitro embryogenic systems.

These embryogenic masses can be subcultured in media with reduced levels of growth regulators (Botti and Vasil, 1984; Goh et al. 2001; Quiroz-Figueroa et al. 2006; Ree and Guerra, 2015) or devoid of them altogether (Vries et al. 1988; Kawahara and Komamine, 1995; Ree and Guerra, 2015; Quiroz-Figueroa et al. 2006). The reduction or removal of regulators inhibits proliferation and promotes a polar auxin gradient (Jiménez, 2001, 2005), stimulating the formation of somatic embryos and their initial development (Kawahara and Komamine, 1995; Von Arnold et al. 2002; Su et al. 2009). Fehér (2006) hypothesizes that the embryogenic determination is achieved in the presence of exogenous auxin. However, its continuous presence blocks the subsequent development of the somatic embryo, i.e., the determination of the path of somatic embryogenesis occurs before the removal of the growth regulator from the culture medium.

This approach (reduction or removal of the growth regulator) is defined by some authors as the somatic embryo differentiation phase, and has been successfully used in *B. gasipaes* (Steinmacher et al. 2007), *E.*

guineensis (Balzon et al. 2013; Jayanthi et al. 2015), and *A. aculeata* (Luis and Scherwinski-Pereira, 2014).

Once differentiated, somatic embryos then undergo a maturation process, commonly characterized by biochemical and morphological changes (Von Arnold et al. 2002). This phase of maturation, marked by rapid expansion of the cotyledons and synthesis and accumulation of reserve compounds, is considered crucial for the successful conversion of somatic embryos into seedlings (Mishra et al. 2012).

Thus, an insufficient maturation process, usually characterized by early germination, leads to the development of underdeveloped seedlings (Jiménez, 2005). To overcome this common problem in somatic embryos of different plant families, researchers have adopted the supplementation of media with growth regulators, such as ABA (Von Arnold et al. 2002; Jiménez, 2005; Othmani et al. 2009; Rai et al. 2011; Fraga et al. 2016b).

It is also emphasized that the accumulation of reserve substances is a typical stage of embryonic development, and is an excellent marker of the quality and fidelity of somatic embryogenesis (Merkle et al. 1995). However, somatic embryos from many palm species show deficient production of some of these reserve compounds, such as proteins (Sané et al. 2006; Aberlenc-Bertossi et al. 2008; Sghaier et al. 2008; Moura et al. 2010). This abnormal protein production, according to Merkle et al. (1995), is a consequence of the types of strategies for maturation used during somatic embryo development and not a problem inherent in somatic embryogenesis. This issue raises the need for optimization of methodologies for palm maturation, which remain little studied (Ree and Guerra, 2015).

The stages after maturation consist of germination, defined as the development of the root system and/or aerial part (Merkle et al. 1995); conversion, conceptualized as the development of seedlings from germinated somatic embryos, (Venkatachalam et al. 2003) and acclimatization, characterized as the transition of plants from *in vitro* to *ex vitro* conditions (Badr et al. 2015). According to Maruyama and Hosoi

(2016) and Guan et al. (2016), the maturation of considerable quantities of somatic embryos and the efficiency of their conversion are considered limiting factors for the use of the technique in many species.

During the germination and conversion phases, there are reports of not only the use of growth regulators such as 3-indoleacetic acid (IAA), indolebutyric acid (AIB), 1-naphthaleneacetic acid (NAA), gibberellic acid (GA_3), and 6-benzylaminopurine (BAP), but also the absence thereof in different species of palm trees. In *E. guineensis*, for example, Jayanthi et al. (2015) and Gomes et al. (2015) reported the use of regulators for rooting purposes. Root formation is considered an essential aspect of the conversion process, since it supplies the water and nutritional needs of plants after transfer to *ex vitro* conditions (Ree and Guerra, 2015). Perera et al. (2009) considered the addition of AG_3 a crucial step for the regeneration of *C. nucifera*, similar to that reported by Freitas et al. (2016), who added BAP, in addition to GA_3, during the regeneration of *E. oleracea*. On the other hand, growth regulators were not required in *E. oleracea* (Scherwinski-Pereira et al. 2012) and *A. aculeata* (Luis and Scherwinski-Pereira, 2014).

Plants transplanted to *ex vitro* conditions should progressively resist increased luminosity (fluxo de fótons), lower relative humidity, temperature change and biotic stresses typical of the new condition (Fki et al. 2011a). Accordingly, different procedures have been adopted to increase the survival and growth of regenerated plants from somatic embryos. Steinmacher et al. (2007), for example, kept plants of *B. gasipaes* inside a plastic box covered with glass to allow light to enter and reduce dehydration. Luis and Scherwinski-Pereira (2014) acclimatized plants of *A. aculeata*, keeping them covered with small transparent plastic bags, with a gradual increase in the opening thereof, until the plants were fully acclimatized to the environmental conditions. It should be noted, however, that as precise and cautious as acclimatization approaches are, according to Von Arnold et al. (2002), the success of survival and growth of regenerated plants depends on the conditions provided in the previous *in vitro* phases.

Figure 3. Phases of somatic embryogenesis in two species of palm trees using zygotic embryos as initial explants (A-C: *Euterpe oleracea* e D-F: *Elaeis guineensis*). A and D: Embryogenic calli originating from the induction phase. B and E: Maturation of somatic embryos. C and F: Regeneration of plants. Bars = A, B, E: 2 mm; C: 1.3 cm; D: 3 mm and F: 1.2 cm. Authorized reproduction - *E. guineensis* (TA Balzon), *E. oleracea* (EO Freitas).

FACTORS THAT INFLUENCE SOMATIC EMBRYOGENESIS

In addition to growth regulators that significantly affect the different phases of somatic embryogenesis, several other aspects influence this technique of *in vitro* culture, including: genotype (Fehér, 2006; Kriaa et al. 2012; Corrêa et al. 2016; Garrocho-Villegas et al. 2017; Ameri et al. 2018; Granja et al. 2018; Ramírez et al. 2018) and physiology of the donor plant (Loyola-Vargas, 2016), type and physiological status of explant development (Sharp et al. 1980; Guerra and Handro, 1991, 1998; Karunaratne et al. 1991; Ledo et al. 2002; Bhatia and Bera, 2015; Bonga, 2016; Isah, 2016), composition of the culture medium (Viñas and Jiménez, 2011; Bhatia and Bera, 2015; Bettencourt et al. 2016; Bonga, 2016; Ameri et al. 2018; Granja et al. 2018), which includes organic molecules excreted by the explant (Nic-Can et al. 2015), pH of the culture medium, (Loyola-

Vargas, 2016) and the duration of the process phases, in addition to the culture conditions (Elmeer, 2013).

It is worth noting that some plant species are less responsive (termo não raramente encontrado na literatura como "recalcitrantes *in vitro*") than others with regard to the initiation of somatic embryogenesis and plant regeneration (Rao, 1996), a fact that encompasses several species (genótipos específicos)of palm trees such as *C. nucifera* (Solís-Ramos et al. 2012; Jayaraj et al. 2015; Nguyen et al. 2015; Karun et al. 2017) and *P. dactylifera* (Al-Khayri, 2011; Sané et al. 2012) and evidences the need for constant optimization of protocols and a better understanding of the molecular barriers that prevent somatic embryogenesis from vegetative tissues(Mozgová et al. 2017). Bonga (2016) mentions that the choice of explants with greater morphogenetic capacity can resolve the issue of recalcitrance in certain plant species.

In relation to the genotype factor, specifically, there are reports evidencing differences in the capability of genotypes of several palms to respond to the various stages of the somatic embryogenesis process (Verdeil et al. 1994; Karun et al. 2004; Steinmacher et al. 2007; Al-Khayri, 2011; Aslam et al. 2011; Sané et al. 2012), as mentioned by Silva et al. (2012) in *E. guineensis*. These latter authors emphasize the importance of the screening for genotypes that are more responsive to the induction of somatic embryogenesis prior to performing basic propagation and regeneration studies.

Regarding for explants used in palm trees, several types are mentioned, such as: zygotic embryos (Guerra and Handro, 1988, 1991, 1998; Goh et al. 2001; Ledo et al. 2002; Steinmacher et al. 2007a; Scherwinski-Pereira et al. 2012; Balzon et al. 2013; Luis and Scherwinski-Pereira, 2014; Freitas et al. 2016; Syuhada et al. 2016; Luna et al. 2017); individual tissues from zygotic embryos, such as plumula (Chan et al. 1998; Azpeitia et al. 2003; Fernando et al. 2003; Pérez-Núñez et al. 2006; Saénz et al. 2010; Rajesh et al. 2014; Jayaraj et al. 2015) and cotyledonary nodes (Jayanthi et al. 2011); leaf tissues (Ahée et al. 1981; Goh et al. 2001; Fki et al. 2003; Karun et al. 2004; Yusnita and Hapsoro, 2011; Kurup et al. 2014; Constantin et al. 2015; Wiendi et al. 2015; Mazri et al. 2017); stem apices - palm trees that produce tillers (Eke et al. 2005; Aslam et al. 2011); root apices (Goh et al. 1999; Goh et al. 2001; Wang et al. 2006; Roshanfekrrad et al. 2017); inflorescences (Guerra and Handro, 1991, 1998; Verdeil et al. 1989; Bhaskaran and Smith, 1992; Fki et al. 2003; Karun et al. 2004; Abul-Soad and Mahdi, 2010; Guedes et al. 2011; Abul-Soad, 2012; Sidky and Eldawyati, 2012; Jayanthi et al. 2015; Jatoi et al. 2015; Sandoval-Cancino et al. 2016; Zayed and Elbar, 2015; Al-Ali et al. 2017; Pádua et al. 2017), flowers (Kriaa et al. 2012; Zayed et al. 2016) and unfertilized ovaries (Perera et al. 2007; Perera et al. 2009).

According to Gaj (2004), the type of explant and its stage of development are probably the most important factors in determining the embryogenic capacity of a culture. The size, origin and age of the explant determine its totipotency and/or multipotency *in vitro*, under controlled environmental conditions, i.e., determine its regeneration capacity (Silva and Dobránszki, 2014). According to Isah (2016), the potential for induction and obtainment of an expressive number of somatic embryos is inversely proportional to the degree of maturity of the explant used. Thus, immature explants are considered ideal, especially for monocotyledonous species (Fehér, 2006).

Among the aforementioned explants, zygotic embryos stand out as the most commonly used explants, primarily because they present the following advantages: high responsiveness to *in vitro* culture, most likely because they present many genes expressed during zygotic embryogenesis that are still active; (Isah, 2016); they are generally free of endophytes and pathogens; they are abundant in many species, and harvesting causes little harm to the mother plant (Ree and Guerra, 2015). However, protocols of regeneration from zygotic embryos present the disadvantage of unpredictability of the nature of the regenerants (Jayanthi et al. 2015), given the cross-pollination (Ree and Guerra, 2015).

We can also mention that examples of success during the processes of induction of somatic embryogenesis and regeneration of palm plants, from explants with a lower degree of maturity, have been common in the literature. Guerra and Handro (1988, 1991, 1998), Gallo-Meagher and Green (2002), Dias et al. (1994) and Scherwinski-Pereira et al. (2012), for example, have reported good results from immature zygotic embryos of *E. edulis*, *S. repens*, *G. gamiova* and *E. oleracea*, respectively, while Jayanthi et al. (2015) reported good results from immature inflorescences of *E. guineensis*.

Immature inflorescences have been projected as promising sources of explants, mainly due to the high embryogenic capacity of originated calli (Fki et al. 2011b), aside from the additional advantages of presenting low fungal and bacterial (Teixeira et al. 1994; Abul-Soad, 2011) contamination, and harvesting with little damage to the mother plant (Teixeira et al. 1994; Jayanthi et al. 2015), depending on the previous knowledge of the arrangement of the inflorescences on the mother plant and the desired degree of maturity of the spathes to be collected. Furthermore, use of these explants makes it possible to clone the source material (Ree and Guerra, 2015), and they exhibit abundant production of inflorescences by many species.

In palm trees, the sectioning of explants, such as young leaves and apical meristems, has been reported with some degree of success, applying the Thin Cell Layers technique (TCL), as mentioned in *E. guineensis* (Scherwinski-Pereira et al. 2010; Guedes et al. 2011; Bonetti et al. 2016), *B. gasipaes* (Steinmacher et al. 2007b; Bettencourt et al. 2016) and *A. aculeata* (Padilha et al. 2015).

Regarding the type of culture medium, according to Viñas and Jiménez (2011), the solid medium has been the most widely used in all stages of the protocols of somatic embryogenesis in palm trees, although there are several reports of the use of liquid medium in *P. dactylifera* (Zouine et al. 2005; Sané et al. 2006; Boufis et al. 2014; Naik and Al-Khayri, 2016) and in *E. guineensis* (Teixeira et al. 1995; Steinmacher et al. 2007; Roowi et al. 2010). The liquid medium has emerged as a promising alternative to *in vitro* propagation, given the additional advantages of uniformity in the production of somatic embryos (Sumaryono et al. 2008), reduced costs for plant production, and the possibility of automation (Neumann et al. 2009).

Additionally, the temporary immersion system, in which cultures are exposed to the liquid medium at certain time intervals (Ree and Guerra, 2015; Monja-Mio et al. 2016), has been successfully employed in the proliferation of embryogenic calli of *E. guineensis* (Marbun et al. 2015) and in the regeneration of *B. gasipaes* (Steinmacher et al. 2011) and *E. guineensis* plants (Gomes et al. 2016).

Among the different basal media formulations, the MS medium (Murashige and Skoog, 1962), according to Ree and Guerra (2015), is the most frequently used in studies on palm trees. Nonetheless, the Y3 medium established by Eeuwens in 1976, specifically for the cultivation of *C. nucifera*, has also been worthy of highlight (Teixeira et al. 1994; Luis and Scherwinski-Pereira, 2014; Jayanthi et al. 2015; Nguyen et al. 2015; Granja et al. 2018). Added to the basal media, sucrose (30 $g.L^{-1}$) and activated charcoal stand out as the carbon source and antioxidant commonly used in protocols of somatic embryogenesis in different species of palm trees (Ree and Guerra, 2015; Saleh and Scherwinski-Pereira et al. 2016).

APPLICATIONS AND BOTTLENECKS OF SOMATIC EMBRYOGENESIS IN PALM TREES

The most attractive application of somatic embryogenesis is the possibility of massive propagation of commercial crops (Von Arnold et al. 2002), especially of several species of palms, due to the difficulties of propagation by conventional methods and the need for clonal propagation of superior genotypes.

Mass production through somatic embryogenesis is already considered a reality in some palm species, such as *P. dactylifera*. Abul-Soad and Mahdi (2010), from a single inflorescence explant of this palm, produced 1,212 rooted plants over a period of two years. Fki et al. (2011b) reported a production rate of 10,000 ± 45 somatic embryos of *P. dactylifera* per liter per month, using suspension cultures. Around the world, specifically in England, France, Israel, Morocco, Oman, United Arab Emirates and Saudi Arabia, there are some tissue culture laboratories that use – among other techniques – somatic embryogenesis for the propagation of *P. dactylifera*, however in a still limited number according to Zaid et al. (2011). A considerably high production of somatic embryos was also achieved in *C. nucifera*, 98,000 somatic embryos/explant, in a protocol developed by Pérez-Núñez et al. (2006), which – according to the authors – represents important progress towards the practical application of the technique.

Due to the production of genetically identical genotypes, somatic embryogenesis has also made it possible to reduce the cycles of genetic improvement of heterozygous plants and plants with long life cycles (Horstman et al. 2017). In *E. guineensis*, for example, vegetative propagation has accelerated and captured the maximum potential yield of selected genotypes, providing an increase of up to 20% when compared to production via seeds (Kushairi et al. 2010).

The technique still stands out due to the possibility of automation during the production and planting in the field, by the large-scale production of somatic embryos in bioreactors (Monja-Mio et al. 2016; Fei and Weathers, 2016) and their encapsulation to produce artificial seeds. Gomes et al. (2016), in *E. guineensis*, developed an efficient protocol using temporary-immersion bioreactors that significantly increased the number of regenerated plants and the formation of secondary embryos. Regarding the production of synthetic seeds, research has been carried out on *P. dactylifera* (Bekheet et al. 2005; Fki et al. 2011b) and *E. guineensis* (Mariani et al. 2014; Palanyandy et al. 2015).

Additionally, as an initial basis for genetic and cellular engineering, somatic embryogenesis plays a significant role in the production of transgenic plants (Quiroz-Figueroa et al. 2006; Wu et al. 2015; Winkelmann, 2016), somatic hybrids and mutants (Gaj, 2004), and is projected as a potential system model for the study of the regulation of gene expression required in the initial stages of higher plant development (Zimmerman, 1993). Embryogenic calli of *P. dactilyfera*, for example, have been used to obtain mutants *in vitro*, and in experiments aimed at genetic transformation in the species (Fki et al. 2011b).

Somatic embryogenesis also supports germplasm preservation techniques, providing cryopreserved material (Von Arnold et al. 2002; Krajňáková et al. 2013; Barra-Jiménez et al. 2015; Winkelmann, 2016), especially proembriogenic masses and somatic embryos (Kulus, 2016), which has been also reported in palm trees (Khawnium and Te-Chato, 2011; Suranthran et al. 2012). Recently, Salma and Engelmann (2017) developed a protocol for long-term cryopreservation of proembriogenic masses of two *P. dactilyfera* cultivars.

On the other hand, it is noteworthy that somatic embryogenesis presents some limitations inherent to the technique, namely: i) asynchronous embryo production (Zegzouti et al. 2001; Shahzad et al. 2016), also reported in different palm species (Ledo et al. 2002; Karun et al. 2004; Steinmacher et al. 2007; Scherwinski-Pereira et al. 2010; Balzon et al. 2013; Luis and Scherwinski-Pereira, 2014; El Bar and El Dawayati, 2014; Pádua et al. 2017), ii) low plant conversion rate, which limits the number of viable plants produced (Nguyen et al. 2015; Karun et al. 2017), (iii) recalcitrance of some plant species (Shahzad et al. 2016; Horstman et al. 2017), and (iv) occurrence of mutations (variações somaclonais) (Fki et al. 2011b; Shahzad et al. 2016; Horstman et al. 2017) that lead to the generation of unpredictable genotypes. It should also be noted, however, that the occurrence of somaclonal changes can be considered a promising source of new and useful varieties (Bhatia and Bera, 2015).

Recalcitrance, for example, is a striking feature of *C. nucifera* (Bhavyashree et al. 2015; Nguyen et al. 2015; Karun et al. 2017). The limited regeneration of *C. nucifera* plants from *in vitro* systems, such as somatic embryogenesis, is still considered a bottleneck for research groups around the world (Nguyen et al. 2015; Karun et al. 2017). Aside from the low conversion rate, the genotypic influence on the responses to *in vitro* cultivation and the production of abnormal somatic embryos are considered limitations to the commercial application of somatic embryogenesis in this palm (Karun et al. 2017).

Plants regenerated from somatic embryos may present somaclonal change, most likely due to the epigenetic (De-La-Peña et al. 2015; Kumar and Staden, 2017) changes that lead to the formation of genotypes with new characteristics, whether desirable or not. An undesirable phenotype, known as a 'mantled' phenotype, characterized by abnormal flower development, has been reported in low percentages in regenerated plants of

E. guineensis, which has been attributed by some authors to the DNA methylation process (Jaligot et al. 2000; Jaligot et al. 2002; Shearman et al. 2013; Ong-Abdullah et al. 2015). Abnormalities in plants regenerated from somatic embryos have also been mentioned in *P. dactylifera* (Al-Mazroui et al. 2007; Bouhouche et al. 2007). Among these abnormalities are female infertility and dwarfism (Ferry, 2011).

Conclusion

Somatic embryogenesis has huge potential for the propagation of ornamental palm trees. This technology can be used for mass propagation, to accelerate breeding programs, and to produce synthetic seeds. Despite the exhaustive literature on the technique and the various protocols developed in different ornamental palms, different bottlenecks need to be overcome for their effective application in commercial scale, among which we can mention the following: (1) reduction of time for obtaining embryogenic calli; (2) maintenance of the embryogenic potential of callogenic formations for longer periods; (3) identification of anatomical, histochemical, ultrastructural, biochemical and molecular markers of the embryogenic potential of callus lineages; (4) increase in the germination frequency of somatic embryos through optimization of the maturation stage and (5) optimization of the acclimatization process of the regenerated plants. Additionally, aspects related to the somatic-embryogenic cell transition need to be better elucidated in palm trees, in order to accelerate the obtainment of quality somatic embryos.

References

Aberlenc-Bertossi, F., Chabrillange, N., Duval, Y., and Tregear, J. (2008). Contrasting globulin and cysteine proteinase gene expression patterns reveal fundamental developmental differences between zygotic and somatic embryos of oil palm. *Tree Physiology*, 28, 1157-1167.

Abul-Soad, A. A. (2012). Influence of inflorescence explant age and 2,4-D incubation period on somatic embryogenesis of date palm. *Emirates Journal of Food and Agriculture*, 24 (5), 434-443.

Abul-Soad, A. A. (2011). Micropropagation of date palm using inflorescence explants. In S.M. Jain, J.M. Al-Khayri, and D.V. Johnson (Eds.), *Date Palm Biotechnology* (pp. 91-118). Dordrecht: Springer.

Abul-Soad, A. A., and Mahdi, S. M. (2010). Commercial production of tissue culture date palm (*Phoenix dactylifera* L.) by inflorescence technique. *Journal of Genetic Engineering and Biotechnology*, 8, 39-44.

Adkins, S. W., Samosir, Y. M., Ernawati, A., Godwin, I. D., and Drew, R. A. (1998). Control of ethylene and use of polyamines can optimise the conditions for somatic embryogenesis in coconut (*Cocos nucifera* L.) and papaya (*Carica papaya* L.). *Acta Horticulturae*, 461, 459-466.

Ahée, J., Arthuis, P., Cas, G., Duval, Y., Guénin, G., Hanower, J., Hanower, P., Lievoux, D., Lioret, C., Malaurie, B., Pannetier, C., Raillot, D., Varechon, C., and Zuckerman, L. (1981). La multiplication végétative *in vitro* du palmier à huile par embryogenèse somatique. *Oléagineux*, 36 (3), 113 -118. [*In vitro* vegetative propagation of oil palm by somatic embryogenesis. *Oleaginous*]

Ahmadi, B., Shariatpanahi, M. E., Ojaghkandi, M. A., and Heydari, A. A. (2014). Improved microspore embryogenesis induction and plantlet regeneration using putrescine, cefotaxime and vancomycin in *Brassica napus* L. *Plant Cell, Tissue and Organ Culture*, 118 (3), 497-505.

Al-Ali, A. M., Ko, C.-Y., Al-Sulaiman, S. A., Al-Otaibi, S. O., Al-Khamees, A. U. H., and Ammar, M. H. (2017). Indirect somatic embryogenesis from mature inflorescence explants of date palm. In J. M. Al-Khayri, S. M. Jain, and D. V. Johnson (Eds.), *Date Palm Biotechnology Protocols. Tissue Culture Applications* (pp. 89-97). Humana Press: New York.

Ali, A-K. A., and Ali, A-K. S. (2015). Effect of different combinations of growth hormones and its interaction on callogenesis. *Research Journal of Biotechnology*, 10 (11), 83-88.

Al-Khayri, J. M. (2011). Basal salt requirements differ according to culture stage and cultivar in date palm somatic embryogenesis. *American Journal of Biochemistry and Biotechnology*, 7 (1), 32-42.

Al-Khayri, J. M., and Ibraheem, Y. (2014). *In vitro* selection of abiotic stress tolerant date palm (*Phoenix dactylifera* L.): A review. *Emirates Journal of Food and Agriculture*, 26 (11), 921-933.

Al-Mazroui, H. S., Zaid, A., and Bouhouche, N. (2007). Morphological Abnormalities in Tissue Culture-Derived Date Palm (*Phoenix dactylifera* L.). *Acta Horticulturae*, 736, 329-335.

Almeida, M., and Almeida, C. V. (2006). Somatic embryogenesis and *in vitro* plant regeneration from pejibaye adult plant leaf primordial. *Pesquisa Agropecuária Brasileira*, 41 (9), 1449-1452.

Almeida, M., Almeida, C. V., Graner, E. M., Brondani, G. E., and Abreu-Tarazi, M. F. (2012). Pre-procambial cells are niches for pluripotent and totipotent stem-like cells for organogenesis and somatic embryogenesis in the peach palm: a histological study. *Plant Cell Reports*, 31, 1495-1515.

Altamura, M. M., Rovere, F. D., Fattorini, L., D'angeli, S., and Falasca, G. (2016). Recent advances on genetic and physiological bases of *in vitro* somatic embryo formation. In M.A. Germanà and M. Lambardi (Eds.), *In vitro embryogenesis in higher plants - Methods in molecular biology* (pp. 47-85). New York: Springer.

Ameri, A., Davarynejad, G. H., Moshtaghi, N., and Tehranifar, A. (2018). Polyethylene glycol and chilling overcome Somatic embryogenesis obstacle in *Pyrus communis*. *Scientia Horticulturae*, 227, 57-64.

Araújo, J. S. P, and Silva, A. M. S. (2010). A palmeira imperial: da introdução no Brasil-Colônia às doenças e pragas no século XXI. *Ciência e Cultura*, 62 (1), 26-28. [The imperial palm: from the introduction in Brazil-Cologne to diseases and plagues in the 21st century. *Science and Culture*]

Aslam, J., Khan, S. A., Cheruth, A. J., Mujib, A., Sharma, M. P., and Srivastava, P. S. (2011). Somatic embryogenesis, scanning electron microscopy, histology and biochemical analysis at different developing stages of embryogenesis in six date palm (*Phoenix dactylifera* L.) cultivars. *Saudi Journal of Biological Sciences*, 18, 369-380.

Azpeitia, A., Chan, J. L., Sáenz, L., and Oropeza, C. (2003). Effect of 22(S),23(S)-homobrassinolide on somatic embryogenesis in plumule explants of *Cocos nucifera* (L.) cultured *in vitro*. *Journal of Horticultural Science & Biotechnology*, 78 (5), 591-596.

Badr, A., Angers, P., and Desjardins, Y. (2015). Comprehensive analysis of *in vitro* to *ex vitro* transition of tissue cultured potato plantlets grown with or without sucrose using metabolic profiling technique. *Plant Cell, Tissue and Organ Culture*, 122, 491-50.

Bhavyashree, U., Jayaraj, K. L., Rachana, K. E., Muralikrishna, K. S., Sajini, K. K., Rajesh, M. K., and Karun, A. (2015). Maintenance of embryogenic potential of calli derived from shoot meristem of West Coast Tall cv. of coconut (*Cocos nucifera* L.). *Journal of Plantation Crops*, 43, 105-116.

Bai, B., Su, Y. H., Yuan, J., and Zhang, X. S. (2013). Induction of somatic embryos in *Arabidopsis* requires local YUCCA expression mediated by the down-regulation of ethylene biosynthesis. *Molecular Plant*, 6 (4), 1247-1260.

Balzon, T. A., Luis, Z. G., and Scherwinski-Pereira, J. E. (2013). New approaches to improve the efficiency of somatic embryogenesis in oil palm (*Elaeis guineensis* Jacq.) from mature zygotic embryos. *In vitro Cellular & Developmental Biology - Plant*, 49 (1), 41-50.

Barra-Jiménez, A., Aronenb, T. S., Alegrea, J., and Toribio, M. (2015). Cryopreservation of embryogenic tissues from mature holm oak trees. *Cryobiology*, 70 (3), 217-225.

Bekheet, S. A., Taha, H. S., and El-Bahr, M. K. (2005). Preservation of date palm cultures using encapsulated somatic embryos. *Arab Journal of Biotechnology*, 8, 319-328.

Bettencourt, G. M. F., Zanella, L. B., Quoirin, M. G. G., and Degenhardt-Goldbach, J. (2016). Efeito da fonte de carbono na embriogênese somática em *Bactris gasipaes*. *Pesquisa Florestal Brasileira*, 36 (86), 179-183. [Effect of carbon source on somatic embryogenesis in *Bactris gasipaes. Brazilian Forest Research*]

Bhaskaran, S., and Smith, R. H. (1992). Somatic embryogenesis from shoot tip and immature inflorescence of *Phoenix dactylifera* cv. Barhee. *Plant Cell Reports*, 12, 22-25.

Bhat, S. V., Booth, S. C., Mcgrath, S. G. K., and Dahms, T. E. S. (2015). *Rhizobium leguminosarum* bv. *viciae* 3841 Adapts to 2,4-Dichlorophenoxyacetic Acid with "Auxin-Like" Morphological Changes, Cell Envelope Remodeling and Upregulation of Central Metabolic Pathways. *PLoS One*, 10 (4), 1-21.

Bhatia, S., and Bera, T. (2015). Somatic Embryogenesis and Organogenesis. In S. Bhatia, K. Sharma, R. Dahiya, T. Bera (Eds.), *Modern applications of plant biotechnology in pharmaceutical sciences* (pp. 210-230). Amsterdam: Elsevier.

Bhojwani, S. S., and Dantu, P. K. (2013). Plant Tissue Culture: An Introductory Text. In S. S. Bhojwani and P. K. Dantu (Eds.), *Cellular Totipotency* (pp. 63-74). New Delhi: Springer.

Bonetti, K. A. P., Nesi, J., Quisen, R. C., and Quoirin, M. (2016). Somatic embryogenesis from zygotic embryos and thin cell layers (TCLs) of Brazilian oil palm (*Elaeis guineensis* × *Elaeis oleifera*). *African Journal of Biotechnology*, 15 (37), 2028-2037.

Bonga, J. M. (2016). Can explant choice help resolve recalcitrance problems in *in vitro* propagation, a problem still acute especially for adult conifers? *Trees*, 1-9.

Borji, M., Bouamama-Gzara, B., Chibani, F., Teyssier, C., Ammar, A. B., Mliki, A., Zekri, S., and Ghorbel, A. (2017). Micromorphology, structural and ultrastructural changes during somatic embryogenesis of a Tunisian oat variety (*Avena sativa* L. var 'Meliane'). *Plant Cell, Tissue and Organ Culture*, 1-14.

Botti, C., and Vasil, I. K. (1984). Ontogeny of somatic embryos of *Pennisetum americanum*. II. In cultured immature inflorescences. *Canadian Journal of Botany*, 62, 1629- 1635.

Boufis, N., Khelifi-Slaouia, M., Djillalia, Z., Djamila, Z., Morslia, A., Bernardsc, M. A., Makhzumc, A., and Khelifia, L. (2014). Somatic embryogenesis in date palm (*Phoenix dactylifera* L. cv. Degla Beida). *Scientia Horticulturae*, 172, 135-142.

Bouhouche, N., Al-Mazroui, H. S., and Zaid, A. (2007). Fertilization failure and abnormal fruit set in tissue culture-derived date palm (*Phoenix dactylifera* L.). *Acta Horticulturae*, 736, 329-335.

Broschat, T. K., Elliott, M. L., and Hodel, D. R. (2014). Ornamental Palms: Biology and Horticulture. In J. Janick (Ed.), *Horticultural Reviews* (pp. 1-120). Nova Jersey: John Wiley & Sons.

Buffard-Morel, J., Verdeil, J.-L., and Pannetier, C. (1992). Embryogenèse somatique du cocotier (*Cocos nucifera* L.) à partir d'explants foliaires: étude histologique. *Canadian Journal of Botany*, 70, 735-741.

Cardoso-Silva, I. M. A. (2018). E*mbriogênese somática de gueroba* [*Syagrus oleracea (Mart.) Becc.*] *com ênfase em aspectos anatômicos, histoquímicos e ultraestruturais* (Tese de doutorado em Ciências Florestais). Repositório Institucional da UnB. [*Somatic embryogenesis of gueroba* [*Syagrus oleracea (Mart.) Becc.*] *with emphasis on anatomical, histochemical and ultrastructural aspects* (Doctoral Thesis in Forest Sciences). Institutional Repository of UnB].

Carvalho, M. A. F., Paiva, R., Alves, E., Nogueira, R. C., Stein, V. C., Castro, E. M., Paiva, P. D. O., and Vargas, D. P. (2013). Morphogenetic potential of native passion fruit (*Passiflora gibertii* N.E. Brown.) calli. *Brazilian Journal of Botany*, *36*, 141-151.

Chan, J. L., Saénz, L., Talavera, C., Hornung, R., Robert, M., and Oropeza, C. (1998). Regeneration of coconut (*Cocos nucifera* L.) from plumule explants through somatic embryogenesis. *Plant Cell Reports*, 17, 515-521.

Chao, C. T., and Krueger, R. R. (2007). *The Date Palm (Phoenix dactylifera L.): Overview of Biology, Uses, and Cultivation*. Hort Science, 42, 1077-1082.

Chehmalee, S., and Te-Chato, S. (2008). Induction of somatic embryogenesis and plantlet regeneration from cultured zygotic embryo of oil palm. *Journal of Agricultural Technology*, *4* (2), 137-146.

Cheng, W. H., Zhu, H. G., Tian, W. G., Zhu, S. H., Xiong, X. P., Sun, Y. Q., Zhu, Q. H., and Sun, J. (2016). De novo transcriptome analysis reveals insights into dynamic homeostasis regulation of somatic embryogenesis in upland cotton (*G. hirsutum* L.). *Plant Molecular Biology*, 92 (3), 279-92.

Cheng, W.-H., Wang, F.-L., Cheng, X.-Q., Zhu, Q.-H., Sun, Y.-Q., Zhu, H.-G., and Sun, J. (2015). Polyamine and its metabolite H_2O_2 play a key role in the conversion of embryogenic callus into somatic embryos in upland cotton (*Gossypium hirsutum* L.). *Frontiers in Plant Science*, 6 (1063).

Cheong, Y. H., Chang, H. S., Gupta, R., Wang, X., Zhu, T., and Luan, S. (2002). Transcriptional profiling reveals novel interactions between wounding, pathogen, abiotic stress, and hormonal responses in *Arabidopsis*. *Plant Physiololy*, 129, 661-677.

Constantin, M., Nchu, W. A., Godswill, N.-N., Wiendi, N. M. A., Wachjar, A., and Frank, N. E. G. (2015). Induction of oil palm (*Elaeis guineensis* Jacq. var. Tenera) callogenesis and somatic embryogenesis from young leaf explants. *Journal of Applied Biology & Biotechnology*, 3 (4), 4-10.

Corrêa, T. R., Motoike, S. Y., Andrade, A. P. S., Coser, S. M., Queiroz, V., Granja, M. M. C., Caetano, D. D. N., Peña, C. N. M., and Picoli, E. A. T. (2016). Accelerated *in vitro* propagation of elite oil palm genotypes (*Elaeis guineensis* Jacq.) by substituting cytokinin with putrescine. *African Journal of Biotechnology*, 15 (50), 2767-2775.

Cvikrová, M., Malá, J., Josef, E., Hrubcovzá, M., and Vágner, M. (1998). Abscisic acid, polyamines and phenolic acids in sessile oak somatic embryos in relation to their conversion potential. *Plant Physiology and Biochemistry*, 36 (3), p. 247-255.

Davletova, S., Mészáros, T., Miskolczi, P., Oberschall, A., Török, K., Magyar, Z., Dudits, D., and Deák, M. (2001). Auxin and heat shock activation of a novel member of the calmodulin like domain protein kinase gene family in cultured alfalfa cells. *Journal of Experimental Botany*, 52, 215-221.

De-La-Peña, C., Nic-Can, G. I., Galaz-Ávalos, R. M., Avilez-Montalvo, R., and Loyola-Vargas, V. M. (2015). The role of chromatin modifications in somatic embryogenesis in plants. *Frontiers in Plant Science*, 6 (635).

Dias, A. C., Guerra, M. P., Cordoba, A. S., and Kemper, E. L. (1994). Somatic embryogenesis and plant regeneration in the tissue culture of *Geonoma gamiova* (Arecaceae). *Acta Horticulturae*, 360, 167-172.

Dransfield, J., Uhl, N. W., Asmussen, C. B., Baker, W. J., Harley, M. M., and Lewis, C. E. (2008). Genera Palmarum: the evolution and classification of palms. *Royal Botanical Garden*, Londres: Kew Publishing.

Dubois, T. M., Guedira, J., Dubois, J., and Vasseur, J. (1990). Direct somatic embryogenesis in roots of Cichorium: as callose an early marker? *Annals of Botany*, 65, 539-545.

Dudits, D., Bögre, L., and Györgyey, J. (1991). Molecular and cellular approaches to the analysis of plant embryo development from somatic cells *in vitro*. *Journal of Cell Science*, 99, 475-484.

Dudits, D., Györgyey, J., Bögre, L., and Bako, L. (1995). Molecular biology of somatic embryogenesis. In T. A. Thorpe (Ed.), *In vitro Embryogenesis in Plants* (pp. 267-308). Dordrecht, The Netherlands: Kluwer Academic Publishers.

Eeuwens, C. J. (1976). Mineral requirements for growth and callus initiation of tissue explants Excised from mature coconut palms (*Cocos nucifera*) and cultured *in vitro*. *Physiologia Plantarum*, 36, 23-28.

Eke, C. R., Akomeah, P., and Asemota, O. (2005). Somatic embryogenesis in date palm (*Phoenix dactylifera* L.) from apical meristem tissues from zebiaand loko landraces. *African Journal of Biotechnology*, 4 (3), 244-246.

El Bar, O. H. A., and El Dawayati, M. M. (2014). Histological changes on regeneration *in vitro* culture of date palm (*Phoenix dactylifera*) leaf explants. *Australian Journal of Crop Science*, 8 (6), 848-855.

Elmeer, K. E. S. (2013). Factors regulating somatic embryogenesis in plants. In J. Aslam, P. S. Srivastava, and M. P. Sharma, M. P. (Eds.),

Somatic embryogenesis and gene expression (pp. 56-81). Narosa Publishing House: New Delhi.

Emons, A. M. C. (1994). Somatic embryogenesis: cell biological aspects. *Acta Botanical Neerlands*, 43 (1), 1-14.

Eshraghi, P., Zarghami, R., and Mirabdulbaghi, M. (2005). Somatic embryogenesis in two Iranian date palm cultivars. *African Journal of Biotechnology*, 4, 1309-1312.

Fehér, A. (2015). Somatic embryogenesis – stress-induced remodeling of plant cell fate. *Biochimica et Biophysica Acta* (BBA), 1849 (4), p. 385-402.

Fehér, A. (2006). Why somatic plant cells start to form embryos? In S. Mujib, and J. Samaj (Eds.). *Somatic embryogenesis* (pp. 85-101). Heidelberg: Plant Cell Monographs, Springer.

Fehér, A., Pasternak, T., and Dudits, D. (2003). Transition of somatic plant cells to an embryogenic state. *Plant Cell, Tissue and Organ Culture*, 74, 201-228.

Fei, L., and Weathers, P. (2016). Bioreactors for Plant Embryogenesis and Beyond. In M.A. Germanà, and M. Lambardi (Eds.), *In vitro embryogenesis in higher plants - Methods in molecular biology* (pp. 245-259). New York: Springer.

Fernando, S. C., Verdeil, J. L., Hocher, V., Weerakoon, L. K., and Hirimburegama, K. (2003). Histological analysis of plant regeneration from plumule explants of *Cocos nucifera*. *Plant Cell Tissue and Organ Culture*, 72 (3), 281-284.

Ferry, M. (2011). Potential of date palm micropropagation for improving small farming systems. In S. M. Jain, J. M. Al-Khayri, and D. V. Johnson (Eds.), *Date Palm Biotechnology* (pp. 15-28). Dordrecht: Springer.

Fki, L., Bouaziz, N., Kriaa, W., Benjemaa-Masmoudi, R., Gargouri-Bouzid, R., Rival, A., and Drira, N. (2011a). Multiple bud cultures of 'Barhee' date palm (*Phoenix dactylifera*) and physiological status of regenerated plants. *Journal of Plant Physiology*, 168, 1694-1700.

Fki, L., Masmoudi, R., Drira, N., and Rival, A. (2003). An optimised protocol for plant regeneration from embryogenic suspension cultures of date palm, *Phoenix dactylifera* L., cv. Deglet Nour. *Plant Cell Reports*, 21, 517-524.

Fki, L., Masmoudi, R., Kriaâ, W., Mahjoub, A., Sghaier, B., Mzid, R., Mliki, A., Rival, A., and Drira, N. (2011b). Date Palm Micropropagation via Somatic Embryogenesis. In S. M. Jain, J. M. Al-Khayri, and D. V. Johnson (Eds.), *Date Palm Biotechnology* (pp. 47-68). Dordrecht: Springer.

Fobert, P. A., and Webb, D. T. (1988). Effects of polyamines, polyamine precursors, and polyamine biosynthetic inhibitors on somatic embryogenesis from eggplant (*Solanum melongena*) cotyledons. *Canadian Journal of Botany*, 66 (9), 1734-1742.

Fraga, H. P. F., Vieira, L. N., Heringer, S. A., Puttkammer, C. C., Silveira, V., and Guerra, M. P. (2016a). DNA methylation and proteome profiles of *Araucaria angustifolia* (Bertol.) Kuntze embryogenic cultures as affected by plant growth regulators supplementation. *Plant Cell, Tissue and Organ Culture*, 125 (2), 353-374.

Fraga, H. P. F., Vieira, L. N., Puttkammera, C. C., Santos, H. P., Garighan, J. A., and Guerra, M. P. (2016b). Glutathione and abscisic acid supplementation influences somatic embryo maturation and hormone endogenous levels during somatic embryogenesis in *Podocarpus lambertii* Klotzsch ex Endl. *Plant Science*, 253, 98-106.

Freitas, E. O., Monteiro, T. R., Nogueira, G. F., and Scherwinski-Pereira, J. E. (2016). Somatic embryogenesis from immature and mature zygotic embryos of the açaí palm (*Euterpe oleracea*): Induction of embryogenic cultures, morphoanatomy and its morphological characteristics. *Scientia Horticulturae*, 212, 126-135.

Gabr, M. F., and Tisserat, B. (1985). Propagating palms *in vitro* with special emphasis on the date palm (*Phoenix dactylifera*). *Scientia Horticulturae*, 25, 255-262.

Gaj, M. D. (2004). Factors influencing somatic embryogenesis induction and plant regeneration with particular reference to *Arabidopsis thaliana* (L.) Heynh. *Plant Growth Regulation*, 43, 27-47.

Gallo-Meagher, M, and Green, J. (2002). Somatic embryogenesis and plant regeneration from immature embryos of saw palmetto, an important landscape and medicinal plant. *Plant Cell, Tissue and Organ Culture*, 68, 253-256.

Garrocho-Villegas, V., Aguilar, R., and Jiménez, E. S. (2017). Contribution of the *Zea mays* insulin-like growth factor (ZmIGF) to the embryogenic competence of maize tissue cultures. *In vitro Cellular & Developmental Biology – Plant*, *53* (2), 122-132.

Gliwicka, M., Nowak, K., Balazadeh, S., Mueller-Roeber, B., and Gaj, M. D. (2013). Extensive modulation of the transcription factor transcriptome during somatic embryogenesis in *Arabidopsis thaliana*. *PLoS One*, 8 (7), 1-20.

Goh, D. K. S., Bon, M.-C., Aliotti, F., Escoute, J., Ferrière, N., and Monteuuis, O. (2001). *In vitro* somatic embryogenesis in two major rattan species: *Calamus merrillii* and *Calamus subinermis*. *In vitro Cellular & Developmental Biology - Plant*, 37, 375-381.

Goh, D. K. S., Michaux-Ferrière, N., Monteuuis, O., and Bom, M-C. (1999). Evidence of somatic embryogenesis from root tip explants of the rattan *Calamus manan*. *In vitro Cellular & Developmental Biology - Plant*, 35, 424-427.

Gomes, H. T., Bartos, P. M. C., Balzon, T. A., and Scherwinski-Pereira, J. E. (2016). Regeneration of somatic embryos of oil palm (*Elaeis guineensis*) using temporary immersion bioreactors. *Industrial Crops and Products*, 89, 244-249.

Gomes, H. T., Bartos, P. M. C., and Scherwinski-Pereira, J. E. (2015). Optimizing rooting and survival of oil palm (*Elaeis guineensis*) plantlets derived from somatic embryos. *In vitro Cellular & Developmental Biology - Plant*, 51, 111-117.

Granja, M. M. C., Motoike, S. Y., Andrade, A. P. S., Correa, T. R., Picoli, E. A. T., and Kuki, K. N., (2018). Explant origin and culture media factors drive the somatic embryogenesis response in *Acrocomia aculeata* (Jacq.) Lodd. ex Mart., an emerging oil crop in the tropics. *Industrial Crops & Products*, *117*, 1-12.

Guan, Y., Li, S.-G., Fan, X.-F., and Su, Z-H. (2016). Application of somatic embryogenesis in woody plants. *Frontiers in Plant Science*, 7 (938) 1-12.

Guedes, R. S., Silva, T. L., Luis, Z. G., and Scherwinski-Pereira, J. N. (2011). Initial requirements for embryogenic calluses initiation in thin cell layers explants from immature female oil palm inflorescences. *African Journal of Biotechnology*, 10 (52), 10774-10780.

Guerra, M. P. and Handro, W. (1998). Somatic embryogenesis and plant regeneration in different organs of *Euterpe edulis* Mart. (Palmae): control and structural features. *Journal of Plant Research*, 111, 65-71.

Guerra, M. P., and Handro, W. (1988). Somatic embryogenesis and plant regeneration in embryo cultures of *Euterpe edulis* mart. (palmae). *Plant Cell Reports*, 7, 550-552.

Guerra, M. P. and Handro, W. (1991). Somatic embryogenesis in tissue cultures of Euterpe edulis Mart. (Palmae). In M.R. Ahuja (Ed.), *Woody Plant Biotechnology* (pp. 189-196). New York: P1enwn Press.

Halperin, W. (1966). Alternative morphogenetic events in cell suspensions. *American Journal of Botany*, 53 (5), 443-453.

Halperin, W., and Jensen, W.A. (1967). Ultrastructural changes during growth and embryogenesis in carrot cell cultures. *Journal of Ultrastructure Research*, 18, 428-443.

Hegazy, A. E., and Aboshama, H. M. (2010). An efficient novel pathway discovered in date palm micropropagation. *Acta Horticulturae*, 882, 167-176.

Herrera, J. C., Cortina, H. A., Anthony, F., Prakash, N. S., Lashermes, P., Gaitán, A. L., Cristancho, M. A., Acuña, R., and Lima, D. R. (2011). Coffee (Coffea spp.). In R. J. Singh (Ed.), *Genetic Resources, Chromosome Engineering, and Crop Improvement Series – Medicinal Plants* (pp. 589-640). Boca Raton: CRC Press.

Horstman, A., Bemer, M., and Boutilier, K. (2017). A transcriptional view on somatic embryogenesis. *Wiley Regeneration*, 4 (4), 201-216.

Ibraheem, Y. M., Pinker, I., and Böhme, M. (2012). The effect of sodium chloride-stress on 'Zaghloul' date palm somatic embryogenesis. *Acta Horticulturae*, 961, 367-373.

Ibrahim, M. S. D., Hartati, R. S., Rubiyo, Purwito, A., and Sudarsono (2013). Direct and indirect somatic embryogenesis on Arabica Coffee

(*Coffea arabica*). *Indonesian Journal of Agricultural Science*, 14 (2), 79-86.

Ikeuchi, M., Sugimoto, K., and Iwase, A. (2013). Plant callus: mechanisms of induction and repression. *Plant Cell*, 25 (9), 3159-73.

Isah, T. (2016). Induction of somatic embryogenesis in woody plants. *Acta Physiologiae Plantarum*, 38 (118), 1-22.

Jaligot, E., Beulé, T., and Rival, A. (2002). Methylation-sensitive RFLPs: characterisation of two oil palm markers showing somaclonal variation-associated polymorphism. *Theoretical and Applied Genetics*, 104, 1263-1269.

Jaligot, E., Rival, A., Beulé, T., Dussert, S., and Verdeil, J. L. (2000). Somaclonal variation in oil palm (*Elaeis guineensis* Jacq.): the DNA methylation hypothesis. *Plant Cell Reports*, 19, 684-690.

Jatoi, M. A., Abul-Soad, A. A., Markhand, G. S., and Solangi, N. (2015). Establishment of an efficient protocol for micropropagation of some Pakistani cultivars of date palm (*Phoenix dactylifera* L.) using novel inflorescence explants. *Pakistan Journal of Botany*, 47 (5), 1921-1927.

Jayanthi, M., Mohan, N. M., and Mandal, P. K. (2011). Direct somatic embryogenesis and plantlet regeneration in oil palm. *Journal of Plant Biochemistry and Biotechnology*, 20, 249-251.

Jayanthi, M., Susanthi, B., Nandiganti, M. M., and Mandal, P. K. (2015). *In vitro* somatic embryogenesis and plantlet regeneration from immature male inflorescence of adult dura and tenera palms of *Elaeis guineensis* (Jacq.). *SpringerPlus*, 4 (256), 1-7.

Jayaraj, K. L., Bhavyashree, U., Fayas, T. P., Sajini, K. K., Rajesh, M. K., and Karun, A. (2015). Histological studies of cellular differentiation during somatic embryogenesis of coconut plumule-derived calli. *Journal of Plantation Crops*, 43 (3), 196-203.

Jiménez, V. M. (2005). Involvement of plant hormones and plant growth regulators on *in vitro* somatic embryogenesis. *Plant Growth Regulation*, 47, 91-110.

Jiménez, V. M. (2001). Regulation of *in vitro* somatic embryogenesis with emphasis on the role of endogenous hormones. *Revista Brasileira de Fisiologia Vegetal*, 13 (2), 196-223.

Jin, F. Y., Hu, L. S., Yuan, D. J., Xu, J., Gao, W. H., He, L. R., Yang, X. Y., and Zhang, X. L. (2014). Comparative transcriptome analysis between somatic embryos (SEs) and zygotic embryos in cotton: evidence for stress response functions in SE development. *Plant Biotechnology Journal*, 12, 161-173.

Johnson, D. V. (1998). *Non-wood forest products 10: Tropical palms*. Rome: Food and Agriculture Organization of the United Nations (FAO).

Kackar, A., and Shekhawat, N. S. (2007). Plant regeneration through somatic embryogenesis and polyamine levels in cultures of grasses of Thar Desert. *Journal of Cell and Molecular Biology*, 6 (2), 121-127.

Kamada, H., Tachikawa, Y., Saitou, T., and Harada, H. (1994). Heat stresses induction of carrot somatic embryogenesis. *Plant Tissue Culture Letters*, 11, 229-232.

Kanchanapoom, K., and Domyoas, P. (1999). The origin and development of embryoids in oil palm (*Elaeis guineensis* Jacq) embryo culture. *Science Asia*, 25, 195-202.

Karami, O., Deljou, A., Esna-Ashari, M., and Ostad-Ahmadi, P. (2006). Effect of sucrose concentrations on somatic embryogenesis in carnation (*Dianthus caryophyllus* L.). *Scientia Horticulturae*, 110, 340-344.

Karami, O., and Said, A. (2010). The molecular basis for stress-induced acquisition of somatic embryogenesis. *Molecular Biology Reports*, 37, 2493-2507.

Karun, A., Rajesh, M. K., Sajini, K. K., Muralikrishna, K. S., Meema, M., Shareefa, M., and Thomas, R. J. (2017). Coconut tissue culture: The Indian initiatives, Experiences and Achievements. *Cord*, 33 (2), 14-24.

Karun, A., Siril, E. A., Radha, E., and Parthasarathy, V. A. (2004). Somatic embryogenesis and plantlet regeneration from leaf and inflorescence explants of arecanut (*Areca catechu* L.). *Current Science*, 86, 1623-1628.

Karunaratne, S., and Periyapperuma, K. (1989). Culture of immature embryos of coconut, *Cocos nucifera* L.: callus proliferation and somatic embryogenesis. *Plant Science*, 62, 247-253.

Kawahara, R., and Komamine, A. (1995). Molecular basis of somatic embryogenesis. In Y. P. S. Bajaj (Ed.), *Biotechnology in Agriculture and Forestry: Somatic Embryogenesis and Synthetic Seed* (pp. 30-40). Berlin: Springer – Verlag.

Kevers, C., Gal, N. L., Monteiro, M., Dommes, J., and Gaspar, T. (2000). Somatic embryogenesis of *Panax ginseng* in liquid cultures: a role for polyamines and their metabolic pathways. *Plant Growth Regulation*, 31, 209-214.

Khawnium, T., and Te-Chato, S. (2011). Simple vitrification protocol for cryopreservation of oil palm using embryogenic culture. *Journal of Agriculture Technology*, 7, 519-529.

Kikuchi, A., Sanuki, N., Higashi, K., Koshiba, T., and Kamada, H. (2006). Abscisic acid and stress treatment are essential for the acquisition of embryogenic competence by carrot somatic cells. *Planta*, 223, 637-645.

Krajňáková, J., Bertolini, A., Gömöry, D., Vianello, A., and Häggman, H. (2013). Initiation, long-term cryopreservation, and recovery of *Abies alba* Mill. embryogenic cell lines. *In vitro Cellular & Developmental Biology - Plant*, 49, 560-571.

Kriaa, W., Sghaier, B., Masmoudi, F., Benjemaa, R., and Drira, N. (2012). The date palm (*Phoenix dactylifera* L.) micropropagation using completely mature female flowers. *Comptes Rendus Biologies*, 335, 194-204.

Krishnan, S. R. S., and Siril, E.A. (2017). Auxin and nutritional stress coupled somatic embryogenesis in *Oldenlandia umbellata* L. *Physiology and Molecular Biology of Plants*, 23 (2), 471-475.

Kulus, D. (2016). Application of cryogenic technologies and somatic embryogenesis in the storage and protection of valuable genetic resources of ornamental plants. In A. Mujib, (Ed.), *Somatic Embryogenesis in Ornamentals and Its Applications* (pp. 1-26). New Delhi: Springer.

Kumar, V., and Van Staden, J. (2017). New insights into plant somatic embryogenesis: an epigenetic view. *Acta Physiologiae Plantarum*, 39 (194), 1-17.

Kurczynska, E. U., Potocka, I., Dobrowolska, I., Kulinska-Lukaszek, K., Sala, K., and Wrobel, J. (2012). Cellular markers for somatic embryogenesis. In K. I. Sato (Ed.), *Embryogenesis* (pp. 307-332). InTech.

Kurup, S. S., Aly, M. A. M., Lekshmi, G., and Tawfik, N. H. (2014). Rapid *in vitro* regeneration of date palm (*Phoenix dactylifera* L.) cv. Kheneizi using tender leaf explant. *Emirates Journal of Food and Agriculture*, 26 (6), 539-544.

Kushairi, A., Tarmizi, A. H., Zamzuri, I., Ong-Abdullah, M. Samsul Kamal, R., Ooi, S. E., and Rajanaidu, N. (2010). Production, performance and advances in oil palm tissue culture. *International Seminar on Advances in Oil Palm Tissue Culture*, p. 1-23.

Ledo, A. S., Lameira, O. A., Benbadis, A. K., De Menezes, I. C., Oliveira, M. S. P., and Medeiros Filho, S. (2002). Somatic embryogenesis from zygotic embryos of *Euterpe oleracea* Mart. *Revista Brasileira de Fruticultura*, 24, 601-603.

Leljak-Levanić, D., Mihaljevic, S., and Bauer, N. (2015). Somatic and zygotic embryos share common developmental features at the onset of plant embryogenesis. *Acta Physiologiae Plantarum*, 37 (127), 1-14.

Litz, R. E. and Schaffer, B. (1987). Polyamines in adventitious and somatic embryogenesis in mango (*Mangifera indica* L.). *Journal of Plant Physiology*, 128, 251-258.

Loyola-Vargas, V. M. (2016). The History of Somatic Embryogenesis. In: V.M. Loyola-Vargas, and N. Ochoa-Alejo (Eds.), *Somatic Embryogenesis: Fundamental Aspects and Applications* (pp. 11-22). Springer: Switzerland.

Loyola-Vargas, V. M., and Ochoa-Alejo, N. (2016). Somatic Embryogenesis. An Overview (pp. 1-8). In V. M. Loyola-Vargas and N. Ochoa-Alejo (Eds.), *Somatic Embryogenesis: Fundamental Aspects and Applications*. Springer: Switzerland.

Luis, Z. G., and Scherwinski-Pereira, J. E. (2014). An improved protocol for somatic embryogenesis and plant regeneration in macaw palm (*Acrocomia aculeata*) from mature zygotic embryos. *Plant Cell Tissue and Organ Culture*, 118, 485-496.

Luna, J. R. H., Sánchez Hinojosa, V. M., Idrovo, N. A. S., and Bustamante, J. F. G. (2017). Estudios histológicos del procesode embriogénesis somática en palma Coco Cumbé (*Parajubaea cocoides* Burret) a partir de embriones cigóticos inmaduros. *Cultivos Tropicales*, 38 (2), 36-44. [Histological studies of the process of somatic embryogenesis in Coco Cumbé palm (*Parajubaea cocoides* Burret) from immature zygotic embryos. *Tropical crops*]

Marbun, C. L. M., Toruan-Mathius, N., Reflini, Utomo, C., and Liwang, T. (2015). Micropropagation of embryogenic callus of oil palm (*Elaeis guineensis* Jacq.) using temporary immersion system. *Procedia Chemistry*, 14, 122-129.

Mariani, T. S., Sasmitamiharja, D., Mienanti, D., Latif, S., Ginting, G., and Miyake, H. (2014). Somatic embryogenesis of oil palm (*Elaeis guineensis* Jacq.) for synthetic seed production. *Asian Journal of Applied Sciences*, 2 (3), 358-367.

Maruyama, T. E., and Hosoi, Y. (2016). Somatic Embryogenesis in Japanese Black Pine (*Pinus thunbergii* Parl.) (pp. 27-40). *In* A. Mujib (Eds.), *Somatic Embryogenesis in Ornamentals and Its Applications*. New Delhi: Springer.

Mazri, M. A., Belkoura, I., Meziani, R., Mokhless, B., and Nour, S. (2017). Somatic embryogenesis from bud and leaf explants of date palm (*Phoenix dactylifera* L.) cv. Najda. *3 Biotech*, 7 (58), 1-9.

Merkle, S. A., Parrott, W. A., and Flinn, B. S. (1995). Morphogenic aspects of somatic. In T.A. Thorpe (Ed.), *In vitro embryogenesis in plant* (pp. 155-203). Dordrecht: Kluwer Academic Publishers.

Mihaljevic, S., Radic, S., Bauer, N., Garic, R., Mihaljevic, B., Horvat, G., Leljak-Levanic, D., and Jelaska, S. (2011). Ammonium-related metabolic changes affect somatic embryogenesis in pumpkin (*Cucurbita pepo* L.). *Journal of Plant Physiology*, 168 (16), 1943-1951.

Mishra, S., Sanyal, I., and Amla, D. V. (2012). Changes in protein pattern during different developmental stages of somatic embryos in chickpea. *Biologia Plantarum*, 56 (4), 613-619.

Monja-Mio, K. M., Herrera-Alamillo, M. A., and Robert, M. L. (2016). Somatic Embryogenesis in Temporary Immersion Bioreactors. *In* V.M. Loyola-Vargas and N. Ochoa-Alejo (Eds.), *Somatic Embryogenesis: Fundamental Aspects and Applications* (pp. 435-454). Springer: Switzerland.

Moon, H.-K., Lee, H., Paek, K.-Y., and Park, S-Y. (2015). Osmotic stress and strong 2,4-D shock stimulate somatic-toembryogenic transition in *Kalopanax septemlobus* (Thunb.) Koidz. *Acta Physiologiae Plantarum*, 37 (1710).

Moura, E. F., Motoike, S. Y., Ventrella, M. C., Sá Júnior, A. Q., Ferreira, M. C., and Carvalho, M. (2009). Somatic embryogenesis in macaw palm (*Acrocomia aculeata*) from zygotic embryos. *Scientia Horticulturae*, 119, 447-454.

Moura, E. F., Ventrella, M. C., and Motoike, S. Y. (2010). Anatomy, histochemistry and ultrastructure of seed and somatic embryo of *Acrocomia aculeata* (Arecaceae). *Scientia Agricola*, 67, 399-407.

Mozgová, I., Muñoz-Viana, R., and Henning, L. (2017). PRC2 Represses Hormone-Induced Somatic Embryogenesis in Vegetative Tissue of *Arabidopsis thaliana*. *PLOS Genetics*, 13 (1), 1-27.

Muilu-Mäkelä, R., Vuosku, J., Hamberg, L., Latva-Mäenpää, H., Häggman, H., and Sarjala, T. (2015). Osmotic stress affects polyamine homeostasis and phenolic content in proembryogenic liquid cell cultures of Scots pine. *Plant Cell Tissue and Organ Culture*, 122, 709-726.

Muniswamy, B., and Sreenath, H. L. (1996). Effect of kanamycin on callus induction and somatic embryogenesis in cultured leaf tissues on *Coffea canephora* Pierre (Robusta). *Journal Coffee Research*, 26 (1), 44-51.

Murashige, T., and Skoog, F. (1962). A revised medium for rapid growth and bioassays with tobacco tissue cultures. *Physiologia*, 15, 473-497.

Na, H., Kim, K. W., Kwack, Y., Kim, S. K., and Chun, C. (2007). Comparative anatomy of embryogenic and non-embryogenic calli from *Pimpinella brachycarpa*. *Journal of Plant Biology,* 50 (3), 344-350.

Naik, P. M., and Al-Khayri, J. M. (2016). Somatic embryogenesis of date palm (*Phoenix dactylifera* L.) through cell suspension culture. In S. M.

Jain (Eds.), *Protocols for in vitro cultures and secondary metabolite analysis of aromatic and medicinal plants - Methods in Molecular Biology* (pp. 357-366). New York: Springer.

Namasivayam, P. (2007). Acquisition of embryogenic competence during somatic embryogenesis. *Plant Cell Tissue and Organ Culture*, 90, 1-8.

Nascimento, W. M. O., and Oliveira, M. S. P. (2011). *Produção de Mudas de Tucumanzeiro-do-pará (Astrocaryum vulgare Mart.) por Perfilhos.* Belém: Embrapa. [*Production of Tucumanzeiro-do-para seedlings (Astrocaryum vulgare Mart.) By Perfilhos*]

Nascimento-Gavioli, M. C. A., Cangahuala-Inocente, G. C., Steinmacher, D., Ree, J. F., Steiner, N., and Guerra, M. P. (2017). Physiological and biochemical features of embryogenic and non-embryogenic peach palm (*Bactris gasipaes* Kunth) cultures. *In vitro Cellular & Developmental Biology – Plant*, 1-8.

Neumann, K., Kumar, A., and Imani, J. (2009). Plant Propagation - Meristem Cultures, Somatic Embryogenesis. In K. Neumann, A. Kumar, and J. Imani, *Plant Cell and Tissue Culture - A Tool in Biotechnology: Basics and Application* (pp. 75-137). Berlin: Springer.

Nguyen, Q. T., Bandupriya, H. D. D., Lopez-Villalobos, A., Sisunandar, S., Foale, M., and Adkins, S. W. (2015). Tissue culture and associated biotechnological interventions for the improvement of coconut (*Cocos nucifera* L.): a review. *Planta*, 242, 1059-1076.

Nic-Can, G. I., Avilez-Montalvo, J. R., Aviles-Montalvo, R. N., Márquez-López, R. E., Mellado-Mojica, E., Galaz-Ávalos, R. M., and Loyola-Vargas, V. M. (2016). The relationship between stress and somatic embryogenesis. In V.M. Loyola-Vargas, N. Ochoa-Alejo (Eds.), *Somatic Embryogenesis: Fundamental Aspects and Applications* (pp. 151-170). Springer: Switzerland.

Nic-Can, G. I., Galaz-Ávalos, R. M., De-La-Peña, C., Alcazar-Magaña, A., Wrobel, K., and Loyola-Vargas, V. M. (2015). Somatic embryogenesis: identified factors that lead to embryogenic repression. A case of species of the same genus. *PLOS ONE*, 10 (6), 1-21.

Nic-Can, G. I., and Loyola-Vargas, V. M. (2016). The role of the auxins during somatic embryogenesis. In V. M. Loyola-Vargas, and N.

Ochoa-Alejo (Eds.), *Somatic Embryogenesis: Fundamental Aspects and Applications* (pp. 171-182). Springer: Switzerland.

Nowak, K., and Gaj, M. D. (2016). Stress-related function of bHLH109 in somatic embryo induction in *Arabidopsis*. *Journal of Plant Physiology*, 193, 119-126.

Ong-Abdullah, M., Ordway, J. M., Jiang, N., Ooi, S.-E., Kok, S.-Y., Sarpan, N., Azimi, N., Hashim, A. T. H., Ishak, Z., Rosli, S. K., Malike, F. A., Bakar, N. A. B., Marjuni, M., Abdullah, N., Yaakub, Z., Amiruddin, M. D., Nookiah, R., Singh, R., Low, E.-T. L., Chan, K.-L., Azizi, N., Smith, S. H., Bacher, B., Budiman, M. A., Brunt, A. V., Wischmeyer, C., Beil, M., Hogan, M. Lakey, N., Lim, C.-C., Arulandoo, X., Wong, C.-K., Choo, C.-N., Wong, W.-C., Kwan, Y.-Y., Alwee, S. S. R. S., Sambanthamurthi, R., and Martienssen, R.A. (2015). Loss of Karma transposon methylation underlies the mantled somaclonal variant of oil palm. *Nature*, 525, 533- 537.

Ooi, S.-E., Choo, C.-N., Ishak, Z., and Ong-Abdullah, M. (2012). A candidate auxin-responsive expression marker gene, EgIAA9, for somatic embryogenesis in oil palm (*Elaeis guineensis* Jacq.). *Plant Cell Tissue and Organ Culture*, 110, 201-212.

Othmani, A. C., Bayoudh, C., Drira, N., Marrakchi, M., and Trifi, M. (2009). Somatic embryogenesis and plant regeneration in date palm *Phoenix dactylifera* L., cv. Boufeggous is significantly improved by fine chopping and partial desiccation of embryogenic callus. *Plant Cell, Tissue and Organ Culture*, 97, 71-79.

Padilha, J. H. D., Ribas, L. L. F., Amano, E., and Quoirin, M. (2015). Somatic embryogenesis in *Acrocomia aculeata* Jacq. (Lodd.) ex Mart using the thin cell layer technique. *Acta Botanica Brasilica, 29* (4), 516-523.

Pádua, M. S., Paiva, L. V., Labory, C. R. G., Alves, E., and Stein, V. C. (2013). Induction and characterization of oil palm (*Elaeis guineensis* Jacq.) pro-embryogenic masses. *Anais da Academia Brasileira de Ciências*, 85 (4), 1545-1556.

Pádua, M. S., Santos, R. S., Labory, C. R. G., Stein, V. C., Mendonça, E. G., Alves, E., and Paiva, L. V. (2017). Histodifferentiation of oil palm somatic embryo development at low auxin concentration. *Protoplasma*, 1-11.

Palanyandy, S. R., Gantait, S., Suranthran, P., Sinniah, U. R., and Subramaniam, S. (2015). Storage of encapsulated oil palm polyembryoids: influence of temperature and duration. *In vitro Cellular & Developmental Biology - Plant*, 51 (1), 118-124.

Pasternak, T. P., Prinsen, E., Ayaydin, F., Miskolczi, P., Potters, G., Asard, H., Vanonckelen, H. A.; Dudits, D., and Fehér, A. (2002). The role of auxin, pH, and stress in the activation of embryogenic cell division in leaf protoplast derived cells of alfalfa. *Plant Physiology*, 129 (4), 1807-1819.

Patnaik, D., Mahalakshmi, A., and Khurana, P. (2005). Effect of water stress and heavy metals on induction of somatic embryogenesis in wheat leaf base cultures. *Indian Journal of Experimental Biology*, 43, 740-745.

Paul, A., Mitter, K., and Raychaudhuri, S. S. (2009). Effect of polyamines on *in vitro* somatic embryogenesis in *Momordica charantia* L. *Plant Cell, Tissue and Organ Culture*, 97, 303-311.

Pereira, C., Montalbán, I. A., García-Mendiguren, O., Goicoa, T., Ugarte, M. D., Correia, S., Canhoto, J. M., and Moncaleán, P. (2016). *Pinus halepensis* somatic embryogenesis is affected by the physical and chemical conditions at the initial stages of the process. *Journal of Forest Research*, 21 (3), 143-150.

Perera, P. I. P., Hocher, V., Verdeil, J.-L., Doulbeau, S., Yakandawala, D. M. D., and Weerakoon, L. K. (2007). Unfertilized ovary: a novel explant for coconut (*Cocos nucifera* L.) somatic embryogenesis. *Plant Cell Reports*, 26, 21-28.

Perera, P. I. P., Vidhanaarachchi, V. R. M., Gunathilake, T. R., Yakandawala, D. M. D., Hocher, V., Verdeil, J. L., and Weerakoon, L. K. (2009). Effect of plant growth regulators on ovary culture of coconut (*Cocos nucifera* L.). *Plant Cell Tissue and Organ Culture*, 99, 73-81.

Pérez-Núñez, M. T., Chan, J. L., Sáenz, L., González, T., Verdeil, J. L., and Oropeza, C. (2006). Improved somatic embryogenesis from *Cocos nucifera* (L.) plumule explants. *In vitro Cellular & Developmental Biology – Plant*, 42, 37-43.

Perianez-Rodriguez, J., Manzano, C., and Moreno-Risueno, M. A. (2014). Post-embryonic organogenesis and plant regeneration from tissues: two sides of the same coin? *Frontiers in Plant Science*, 5 (219), 1-11.

Petrek, J., Zitka, O., Adam, V., Bartusek, K., Anjum, N. A., Pereira, E., Havel, L., and Kizek, R. (2015). Are Early Somatic Embryos of the Norway Spruce (*Picea abies* (L.) Karst.) Organised? *PLOS ONE*, 10 (12), 1-16.

Puigderrajols, P., Mir, G., and Molinas, M. (2001). Ultrastructure of early secondary embryogenesis by multicellular and unicellular pathways in cork oak (*Quercus suber* L.). *Annals of Botany*, 87, 179-189.

Purohit, S.D. (2013). Plant Cell Culture: Basic Concepts. In S.D. Purohit, *Introduction to Plant Cell Tissue and Organ Culture*. Delhi: PHI Learning.

Quiroz-Figueroa, F., Rojas-Herrera, R., Galaz-Avalos, R. M., and Loyola-Vargas, V. M. (2006). Embryo production through SE can be used to study cell differentiation in plants. *Plant Cell Tissue Organ Cult*, *86*, 285-301.

Raemakers, C. J. J. M., Jacobsen, E., and Visser, R. G. F. (1995). Secondary somatic embryogenesis and applications in plant breeding. *Euphytica*, 81, 93-107.

Rai, M. K., Shekhawat, N. S., Gupta, A. K., Phulwaria, M., Ram, K., and Jaiswal, U. (2011). The role of abscisic acid in plant tissue culture: a review of recent progress. *Plant Cell, Tissue and Organ Culture*, 106, 179-190.

Rajesh, M. K., Radha, E., Karun, A., and Parthasarathy, V. A. (2003). Plant regeneration from embryo-derived callus of oil palm – the effect of exogenous polyamines. *Plant Cell, Tissue and Organ Culture*, 75, 41-47.

Rajesh, M. K., Radha, E., Sajini, K. K., and Karun, A. (2014). Polyamine-induced somatic embryogenesis and plantlet regeneration *in vitro* from plumular explants of dwarf cultivars of coconut (*Cocos nucifera*). *The Indian Journal of Agricultural Sciences*, 84 (4), 527-530.

Ramírez, A. M. H., Vasquez, T. H., Osorio, T. M. O., Garcés, L. A., and Trujillo, A. I. U. (2018). Evaluation of the potential of regeneration of different Colombian and commercial genotypes of cocoa (*Theobroma cacao* L.) via somatic embryogenesis. *Scientia Horticulturae*, 229, 148-156.

Rao, K. S. (1996). Embryogenesis in flowering plants – recent approaches and prospects. *Journal of Biosciences*, 21, 827-841.

Ree, J. F., and Guerra, M. P. (2015). Palm (Arecaceae) somatic embryogenesis. *In vitro Cellular & Developmental Biology - Plant*, 51 (6), 589-602.

Reinert, J. (1958). Über die kontrolle der morphogenese und die induktion von adventiv embryonene an gewebekulturen aus karotten. [About control of morphogenesis and induction of adventive embryonic tissue cultures from carrots.] *Planta*, 53, 318-333.

Reis, R. S., Vale, E. M., Heringer, A. S., Santa-Catarina, C., and Silveira, V (2016). Putrescine induces somatic embryo development and proteomic changes in embryogenic callus of sugarcane. *Journal of Proteomics*, 130, 170-179.

Rocha, D. I., Kurczyńska, E., Potocka, I., Steinmacher, D. A., and Otoni, W.C. (2016). Histology and Histochemistry of Somatic Embryogenesis. In V. M. Loyola-Vargas, and N. Ochoa-Alejo (Eds.), *Somatic Embryogenesis: Fundamental Aspects and Applications* (pp. 471-494). Springer: Switzerland.

Ronchi, V. N., and Giorgeiti, L. (1995). The Cell's Commitment to Somatic Embryogenesis. In Y. P. S. Bajaj (Ed.), *Biotechnology in Agriculture and Forestry: Somatic Embryogenesis and Synthetic Seed* (pp. 3-19). Berlin: Springer – Verlag.

Roowi, S. H., Ho, C.-L., Alwee, S. S. R. S., Abdullah, M. O., and Napis, S. (2010). Isolation and characterization of differentially expressed transcripts from the suspension cells of oil palm (*Elaeis guineensis* Jacq.) in response to different concentration of auxins. *Molecular Biotechnology*, 46, 1-19.

Roshanfekrrad, M., Zarghami, R., Hassani, H., Zakizadeh, H., and Salari, A. (2017). Effect of AGNO3 and BAP on root as a novel explant in date palm (*Phoenix dactylifera* cv. Medjool) somatic embryogenesis. *Pakistan Journal of Biological Sciences*, 20 (1), 20-27.

Saénz, L., Azpeitia, A., Chuc-Armendariz, B., Chand, J. L., Verdeil, J. L., Hocher, V., and Oropeza, C. (2006). Morphological and histological changes during somatic embryo formation from coconut plumule explants. *In vitro Cellular & Developmental Biology - Plant*, 42, 19-25.

Saénz, L., Herrera-Herrera, G., Uicab-Ballote, F., Chan, J. S., and Oropeza, C. (2010). Influence of form of activated charcoal on embryogenic callus formation in coconut (*Cocos nucifera*). *Plant Cell Tissue and Organ Culture*, 100, 301-308.

Saldanha, C. W., and Martins-Corder, M. P. (2012). *In vitro* germination and embryogenic competence acquisition of *Euterpe edulis* Martius immature zygotic embryos. *Crop Breeding and Applied Biotechnology*, 12, 171-178.

Saleh, E. O. L. (2016). *Fisiologia da germinação in vitro, embriogênese somática e conservação ex situ de babaçu (Attalea speciosa Mart. ex Spreng)* (Tese de Doutorado em Botânica). Repositório Institucional da UnB. [*Physiology of in vitro germination, somatic embryogenesis and ex situ conservation of babassu (Attalea speciosa Mart. ex Spreng)* (Doctoral Thesis in Botany). Institutional Repository of UnB].

Saleh, E. O. L. and Scherwinski-Pereira, J. E. (2016). Advances in somatic embryogenesis of palm trees (Arecaceae): fundamentals and review of protocols. In A. Mujib, (Ed.), *Somatic Embryogenesis in Ornamentals and Its Applications* (pp. 231-254). New Delhi: Springer.

Salma, M., and Engelmann, F. (2017). Cryopreservation of Date Palm Pro-Embryonic Masses Using the D Cryo-plate Technique. In J. Al-Khayri, S. Jain, and D. Johnson (Eds.), *Date Palm Biotechnology Protocols Volume II: Germplasm Conservation and Molecular Breeding, Methods in Molecular Biology* (pp. 25-37). New York: Humana Press.

Samaj, J., Baluska, F., Bobák, M., and Volkmann, D. (1999). Extracellular matrix surface network of embryogenic units of friable maize callus contains arabinogalactan-proteins recognized by monoclonal antibody JIM4. *Plant Cell Reports*, 18, 369-374.

Sandoval-Cancino, G., Sáenz, L., Chan, J. L., and Oropeza, C. (2016). Improved formation of embryogenic callus from coconut immature inflorescence explants. *In vitro Cellular & Developmental Biology – Plant,* 52 (4), 367-378.

Sané, D., Aberlenc-Bertossi, F., Diatta, L. I. D., Guèye, B., Daher, A., Sagna, M., Duval, Y., and Borgel, A. (2012). Influence of Growth Regulators on Callogenesis and Somatic Embryo Development in Date Palm (*Phoenix dactylifera* L.) Sahelian Cultivars. *The Scientific World Journal*, 1-8.

Sané, D., Aberlenc-Bertossi, F., Gassama-Dia, Y. K., Sagna, M., Trouslot, M. F., Duval, Y., and Borgel, A. (2006). Histocytological analysis of callogenesis and somatic embryogenesis from cell suspensions of date palm (*Phoenix dactylifera*). *Annals of Botany*, 98, 301-308.

Santarem, E., Pelissier, B., and Finer, J. (1997). Effect of explant orientation, pH, solidifying agent and wounding on initiation of soybean somatic embryos. *In vitro Cellular & Developmental Biology - Plant*, 33, 13-19.

Sarasan, V., Ramsay, M. M., and Roberts, A. V. (2002). *In vitro* germination and induction of direct somatic embryogenesis in 'Bottle Palm' [*Hyophorbe lagenicaulis* (L. Bailey) H.E. Moore], a critically endangered Mauritian palm. *Plant Cell Reports*, 20, 1107-1111.

Scherwinski-Pereira, J. E., Da Guedes, R. S., Fermino Jr., P. C. P., Silva, T. L., and Costa, F. H. S. (2010). Somatic embryogenesis and plant regeneration in oil palm using the thin cell layer technique. *In vitro Cellular & Developmental Biology - Plant*, 46, 378-385.

Scherwinski-Pereira, J. E., Guedes, R. S., Da Silva, R. A., Fermino Jr., P. C. P., Luis, Z. G., and Freitas, E. O. (2012). Somatic embryogenesis and plant regeneration in açaí palm (*Euterpe oleracea*). *Plant Cell Tissue and Organ Culture*, 109, 501-508.

Sghaier, B., Bahloul, M., Gargouri Bouzid, R., and Drira, N. (2008). Development of zygotic and somatic embryos of *Phoenix dactylifera* L. cv. Deglet Nour: comparative study. *Scientia Horticulturae*, 116, 169-175.

Shahzad, A., Sharma, S., and Siddiqui, S. A. (2016). Somatic Embryogenesis: A Valuable Strategy for Phyto-Climbing Diversity Conservation. In A. Shahzad, S. Sharma, and S. A. Siddiqui (Eds.), *Biotechnological strategies for the conservation of medicinal and ornamental climbers* (pp. 195-216). New York: Springer.

Shang, H.-H., Liu, C.-L., Zhang, C., Li F.-L., Hong, W.-D., and Li, F.-G. (2009). Histological and ultrastructural observation reveals significant cellular differences between *Agrobacterium* transformed embryogenic and non-embryogenic calli of cotton. *Journal of Integrative Plant Biology*, 51, 456-465.

Sharp, W. R., Sondahl, M. R., Caldas, L. S., and Maraffa, S. B. (1980). The physiology of *in vitro* asexual embryogenesis. *Horticultural Reviews*, 2, 268-310.

Shearman, J. R., Jantasuriyarat, C., Sangsrakru, D., Yoocha, T., Vannavichit, A., Tragoonrung, S., and Tangphatsornruang, S. (2013). Transcriptome analysis of normal and mantled developing oil palm flower and fruit. *Genomics*, 101 (5), 306-312.

Sidky, R. A., and Eldawyati, M. M. (2012). Proliferation of female inflorescences explants of date palm. *Annals of Agricultural Science*, 57 (2), 161-165.

Silva, J. A. T., and Dobránszki, J. (2014). Dissecting the concept of the thin cell layer: theoretical basis and practical application of the Plant Growth Correction Factor. *Journal of Plant Growth Regulation*, 33, 881-895.

Silva, R. C., Luis, Z. L., and Scherwinski-Pereira, J. E. (2012). Differential responses to somatic embryogenesis of different genotypes of Brazilian oil palm (*Elaeis guineensis* Jacq.). *Plant Cell Tissue and Organ Culture*, 111, 59-67.

Silveira, V., Vita, A. M., Macedo, A. F., Dias, M. F. R., Floh, E. I. S., and Santa-Catarina, C. (2013). Morphological and polyamine content changes in embryogenic and non-embryogenic callus of sugarcane. *Plant Cell Tissue and Organ Culture*, 114 (3), 351-64.

Sivanesan, I., Lim, M. Y., and Jeong, B. R. (2011). Somatic embryogenesis and plant regeneration from leaf and petiole explants of *Campanula punctata* Lam. var. *rubriflora* Makino. *Plant Cell Tissue and Organ Culture*, 107, 365-369.

Solís-Ramos, L. Y., Andrade-Torres, A., Carbonell, L. A. S., Salín, C. M. O., and Serna, E. C. (2012). Somatic embryogenesis in recalcitrant plants. In K. I. Sato (Ed.), *Embryogenesis* (pp. 597-618). Rijeka: InTech.

Staritsky, G. (1970). Tissue culture of the oil palm (*Elaeis guineensis* Jacq.) as a tool for its vegetative propagation. *Euphytica*, 19, 288-292.

Steinmacher, D. A., Cangahuala-Inocente, G. C., Clement, C. R., and Guerra, M. P. (2007a). Somatic embryogenesis from peach palm zygotic embryos. *In vitro Cellular and Developmental Biology - Plant*, 43, p. 124-132.

Steinmacher, D. A., Clement C. R., and Guerra, M. P. (2007). Somatic embryogenesis from immature peach palm inflorescence explants: towards development of an efficient protocol. *Plant Cell Tissue and Organ Culture*, 89, 15-22.

Steinmacher, D. A., Guerra, M. P., Saare-Surminski, K., and Lieberei, R. (2011). A temporary immersion system improves *in vitro* regeneration of peach palm through secondary somatic embryogenesis. *Annals of Botany*, 108 (8), 1463-1475.

Steinmacher, D. A., Krohn, N. G., Dantas, A. C. M., Stefenon, V. M., Clement, C. R., and Guerra, M. P. (2007b). Somatic embryogenesis in peach palm using the thin cell layer technique: Induction, morphohistological aspects and AFLP analysis of somaclonal variation. *Annals of Botany*, 100, 699-709.

Steward, F. C., Mapes M. O., and Mears, K. (1958). Growth and organized development of cultured cells. II. Organization in culture grown freely suspended cells. *American Journal of Botany*, 45, 705-708.

Su, Y. H., Zhao, X. Y., Liu, Y. B., Zhang, C. L., O'neill, S. D., and Zhang, X. S. (2009). Auxin-induced WUS expression is essential for embryonic stem cell renewal during somatic embryogenesis in *Arabidopsis*. *The Plant Journal*, 59, 448-460.

Sumaryono, Riyadia, I., Kasia, P. D., and Ginting, G. (2008). Growth and differentiation of embryogenic callus and somatic embryos of oil palm

(*Elaeis guineensis* jacq.) in temporary immersion system. *Indonesian Journal of Agriculture*, 1 (2), 109-114.

Suranthran, P., Gantait, S., Sinniah, U. R., Subramaniam, S., Alwee, S. S. R. S., and Roowi, S. H. (2012). Effect of loading and vitrification solutions on survival of cryopreserved oil palm polyembryoids. *Plant Growth Regulation*, 66, 101-109.

Syuhada, W. N., Rasid, O. A., and Parveez, G. K. A. (2016). Evaluation on the effects of culture medium on regeneration of oil palm plantlets from immature embryos (ie). *Journal of Oil Palm Research*, 28 (2), 234-239.

Tagipur, E. M., Seker, G., Silva, J. A. T., and Mendi, Y. Y. (2016). Somatic embryogenesis, cryopreservation, and *in vitro* mutagenesis in Cyclamen. In A. Mujib (Ed.), *Somatic Embryogenesis in Ornamentals and Its Applications* (pp. 155-168). New Delhi: Springer.

Taha, R. A., and Hassan, M. M. (2014). Using low levels of seawater to enhance growth and development of date palm embryogenic cultures. *Asian Journal of Agricultural Sciences*, 6 (2), 69-74.

Takeda, T., Hayakawa, F., Oe, K., and Matsuoka, H. (2002). Effects of exogenous polyamines on embryogenic carrot cells. *Biochemical Engineering Journal*, 12, 21-28.

Talapatra, S., Goswami, P., Das, S., and Raychaudhuri, S. S. (2016). Role of SERK during somatic embryogenesis and its interaction with brassinosteroids. In A. Mujib (Ed.), *Somatic Embryogenesis in Ornamentals and Its Applications* (pp. 141-154). New Delhi: Springer.

Tan, Y., Hu, W., Xu, X., Zhou, J., Wang, C., Liu, X., and Cheng, G. (2017). Polyamine plays a role in subculture growth of *in vitro* callus of Indica rice. *Acta Biologica Cracoviensia Series Botanica*, 59 (1).

Tao, L., Yang, Y., Wang, Q., and You, X. (2012). Callose deposition is required for somatic embryogenesis in plasmolyzed *Eleutherococcus senticosus* zygotic embryos. *International Journal of Molecular Sciences*, 13, 14115-14126.

Te-Chato, S., and Hilae, A. (2007). High-frequency plant regeneration through secondary somatic embryogenesis in oil palm (*Elaeis guineensis* Jacq. var. tenera). *Journal of Agricultural Technology*, 3 (2), 345-357.

Teixeira, J. B., Söndahl, M. R., and Kirby, E. G. (1994). Somatic embryogenesis from immature inflorescences of oil palm. *Plant Cell Reports*, 13, 247-250.

Teixeira, J. B., Söndahl, M. R., and Kirby, E. G. (1993). Somatic embryogenesis from immature zygotic embryos of oil palm. *Plant Cell Tissue and Organ Culture*, 34, 227-233.

Teixeira, J. B., Söndahl, M. R., Nakamura, T., and Kirby, E. G. (1995). Establishment of oil palm cell suspensions and plant regeneration. *Plant Cell Tissue and Organ Culture*, 40, 105-111.

Thuzar, M., Vanavichit, A., Tragoonrung, S., and Jantasuriyarat, C. (2011). Efficient and rapid plant regeneration of oil palm zygotic embryos cv. 'Tenera' through somatic embryogenesis. *Acta Physiologiae Plantarum*, 33, 123-128.

Tisserat, B. (1979). Propagation of date palm (*Phoenix dactylifera* L.) *in vitro*. *Journal of Experimental Botany*, 30 (119), 1275-1283.

Tomlinson, P. B., and Huggett, B. A. (2012). Cell longevity and sustained primary growth in palm stems. *American Journal of Botany*, 99 (12), 1891-1902.

Valverde, R., Arias, O., and Thorpe, T. A. (1987). Picloram-induced somatic embryogenesis in pejibaye palm (*Bactris gasipaes* HBK). *Plant Cell Tissue and Organ Culture*, 10, 149-15.

Venkatachalam, P., Geetha, N., Priya, P., Jayabalan, N., and Sita, G. L. (2003). Somatic embryogenesis. In P. K. Jaiwal, and R. P. Singh (Eds.), *Improvement Strategies for Leguminosae Biotechnology* (pp. 87-132). Dordrecht: Kluwer Academic Publishers.

Veramendi, J., and Navarro, L. (1996). Influence of physical of nutrient medium and sucrose on somatic embryogenesis of date palm. *Plant Cell Tissue and Organ Culture*, 45, 159-164.

Verdeil, J. L., Alemanno, L., Niemenak, N., and Trambarger, T. J. (2007). Pluripotent versus totipotent plant stem cells: dependence versus autonomy? *Trends in Plant Science*, 12, 245-252.

Verdeil, J. L., Buffard-Morel, J., and Pannetier, C. (1989). Embryogénèse somatique du cocotier (*Cocos nucifera* L.) à partir de tissus foliaires et inflorescenciels: bilan des recherches et perspectives. *Oléagineux*, 44, 403-411. [Somatic embryogenesis of coconut palms (*Cocos nucifera* L.) from foliar and inflorescence tissues: review of research and perspectives. *Oleaginous*]

Verdeil, J. L., Huet, C., Grosdemange, F., and Buffard-Morel, J. (1994). Plant regeneration from cultured immature inflorescences of coconut (*Cocos nucifera* L.): evidence for somatic embryogenesis. *Plant Cell Reports*, 13, 218-221.

Verdeil, J. L, Hocher, V., Huet, C., Grosdemange, F., Escoute, J., Ferriere, N., and Nicole, M. (2001). Ultrastructural changes in coconut calli associated with the acquisition of embryogenic competence. *Annals of Botany*, 88, 9-18.

Viñas, M., and Jiménez, V. M. (2011). Factores que influyen en la embriogénesis somática *in vitro* de palmas (Arecaceae). [Factors that influence the in vitro somatic embryogenesis of palms (Arecaceae).] *Revista Colombiana de Biotecnologia*, 13, 229-242.

Von Arnold, S., Sabala, I., Bozhkov, P., Dyachok, J., and Filonova, L. (2002). Developmental pathways of somatic embryogenesis. *Plant Cell Tissue and Organ Culture*, 69, 233-249.

Vondráková, Z., Eliášová, K., Vágner, M., Martincová, O., and Cvikrová, M. (2015). Exogenous putrescine affects endogenous polyamine levels and the development of *Picea abies* somatic embryos. *Plant Growth Regulation*, 75, 405-414.

Vries, S. C., Booij, H., Meyerink, P., Huisman, G., Wilde, H. D., Thomas, T. L., and Van Kammen, A. (1988). Acquisition of embryogenic potential in carrot cell-suspension cultures. *Planta*, 176, 196-204.

Wang, H. C., Chen, J. T., and Chang, W. C. (2010). Morphogenetic routes of long-term embryogenic callus culture of *Areca catechu*. *Biologia Plantarum*, 54, 1-5.

Wang, H. C., Chen, J. T., Wu, S. P., Lin, M. C., and Chang, W. C. (2003). Plant regeneration through somatic embryogenesis from zygotic embryo-derived callus of *Areca catechu* L. (Arecaceae). *In vitro Cellular & Developmental Biology – Plant*, 39, 34-36.

Wang, H. C., Chen, J. T., and Chang, W. C. (2006). Somatic embryogenesis and plant regeneration from leaf, root and stem-derived callus cultures of *Areca catechu*. *Biologia Plantarum*, 50 (2), 279-282.

Wang, K.L., Li, H., and Ecker, J. R. (2002). Ethylene biosynthesis and signaling networks. *Plant Cell*, 14, 131-151.

Wang, X., Shi, L., Lin, G., Pan, X., Chen, H., Wu, X., Takác, T., Samaj, J., and Xu, C. (2013). A systematic comparison of embryogenic and non-embryogenic cells of banana (*Musa* spp. AAA): Ultrastructural, biochemical and cell wall component analyses. *Scientia Horticulturae*, 159, 178-185.

Wickramasuriya, A. M., and Dunwell, J. M. (2015). Global scale transcriptome analysis of *Arabidopsis* embryogenesis *in vitro*. *BMC Genomics*, 16 (301), 1-23.

Wiendi, N. M. A., Constantin, M., and Wachyar, A. (2015). Study of Oil Palm (*Elaeis guineensis* Jacq.) *In vitro* Embryogenesis using Young Leaf Explants. *Journal of Tropical Crop Science*, 2 (2), 5-9.

Williams, E. S., and Maheswaran, B. (1986). Somatic embryogenesis: Factors influencing coordinated behaviour of cells as an embryogenic group. *Annals of Botany*, 57, 443-462.

Winkelmann, T. (2016). Somatic versus zygotic embryogenesis: learning from seeds. In M. A. Germanà, and M. Lambardi (Eds.), *In vitro embryogenesis in higher plants - Methods in molecular biology* (pp. 25-46). New York: Springer.

Wu, J., Liu, C., Seng, S., Khan, M.A., Sui, J., Gong, B., Liu, C., Wu, C., Zhong, X., He, J., and Yi, M. (2015). Somatic embryogenesis and *Agrobacterium*-mediated transformation of *Gladiolus hybridus* cv. 'Advance Red'. *Plant Cell Tissue and Organ Culture*, 120, 717-728.

Yang, X. Y., and Zhang, X. L. (2010). Regulation of somatic embryogenesis in higher plants. *Critical Reviews in Plant Sciences*, 29 (1), 36-57.

Yeung, E. C. (1995). Structural and development patterns in somatic embryogenesis. In T.A. Thorpe (Ed.), *In vitro Embryogenesis in Plants* (pp. 205-247). Dordrecht: Kluwer Academic Publishers.

You, X. L., Yi, J. S., and Choi, Y. E. (2006). Cellular change and callose accumulation in zygotic embryos of *Eleutherococcus senticosus* caused by plasmolyzing pretreatment result in high frequency of single-cell-derived somatic embryogenesis. *Protoplasma*, 227, 105-112.

Yusnita and Hapsoro, D. (2011). *In vitro* callus induction and embryogenesis of oil palm (*Elaeis guineensis* Jacq.) from leaf explants. *Hayati Journal of Biosciences*, 18 (2), 61-65.

Zaid, A., El-Korchi, B., and Visser, H. J. (2011). Commercial Date Palm Tissue Culture Procedures and Facility Establishment. In S. M. Jain, J. M. Al-Khayri, and J. Dennis (Eds.), *Date Palm Biotechnology* (pp. 137-180). Netherlands: Springer.

Zavattieri, M. A., Frederico, A. M., Lima, M., Sabino, R., and Arnholdt-Schmitt, B. (2010). Induction of somatic embryogenesis as an example of stress-related plant reactions. *Electronic Journal of Biotechnology*, 13 (1), 1-9.

Zayed, E. M. M., and Abd Elbar, O. H. (2015). Morphogenesis of immature female inflorescences of date palm *in vitro*. *Annals of Agricultural Science*, 60 (1), 113-120.

Zayed, E. M. M., Zein El Din, A. F. M., Manaf, H. H., and Abdelbar, O.H. (2016). Floral reversion of mature inflorescence of date palm *in vitro*. *Annals of Agricultural Science*, 1-9.

Zegzouti, R., Arnould, M.-F., and Favre, J.-M. (2001). Histological investigation of the multiplication steps in secondary somatic embryogenesis of *Quercus robur* L. *Annals of Forest Science*, 58, 681-690.

Zeng, F., Zhang, X., Jin, S., Cheng, L., Liang, S., Hu, L., Guo, X., Nie, Y., and Cao, J. (2007). Chromatin reorganization and endogenous auxin/cytokinin dynamic activity during somatic embryogenesis of cultured cotton cell. *Plant Cell Tissue and Organ Culture*, 90, 63-70.

Zhou, T., Yang, X., Guo, K., Deng, J., Xu, J., Gao, W., Lindsey, K., and Zhang, X. (2016). ROS homeostasis regulates somatic embryogenesis via the regulation of auxin signaling in cotton. *Molecular & Cellular Proteomics*, 15, 2108-2124.

Zhou, X., Zheng, R., Liu, G., Xu, Y., Zhou, Y., Laux, T., Zhen, Y., Harding, S. A., Shi, J., and Chen, J. (2017). Desiccation Treatment And Endogenous IAA Levels Are Key Factors Influencing High Frequency Somatic Embryogenesis In *Cunninghamia lanceolata* (Lamb.) Hook. *Frontiers in Plant Science*, 8 (2054), 1-15.

Zimmerman, J. L. (1993). Somatic embryogenesis: a model for early development in higher plants. *The Plant Cell*, 5, 1411-1423.

Zouine, J., El Bellaj, M., Meddich, A., Verdeil, J.-L., and El Hadrami, I. (2005). Proliferation and germination of somatic embryos from embryogenic suspension cultures in *Phoenix dactylifera*. *Plant Cell Tissue and Organ Culture*, 82, 83-92.

In: Micropropagation of Ornamental Plants ISBN: 978-1-53614-541-0
Editor: Taoufik Bettaieb

Chapter 9

A SHORTENED PROTOCOL FOR THE MICROPROPAGATION OF A BIRD OF PARADISE (*STRELIZIA REGINAE* AIT.)

Sarra Arbaoui, Sihem Soufi and Taoufik Bettaieb*
Laboratory of Horticultural Sciences, National Agronomic Institute of Tunisia, University of Carthage, Tunisia

ABSTRACT

Strelitzia reginae Ait. is one among the most economically prominent ornamental plants prized for their colorful flowers and the high post-harvest durability. Despite of the high demands for these tropical perennial plants they have not been widely spread due to the low rate of multiplication and unavailability of uniform cultivars. Reliable micropropagation methods have not yet been also fully developed to ensure their large scale production. The limited success of these methods is reported to be limited due to oxidative browning of the wounded explants. In this chapter a successful protocol was described for micropropagation of ornamental plant *Strelitzia reginae* using apical buds as source of explants. Apical buds were capable of producing multiple

* Corresponding Author Email: tbettaieb@yahoo.fr.

shoots *in vitro* on MS medium supplemented with 2 mg/l of BA and 0.25 mg/l of IBA. *In vitro* multiplied shoots were successfully *ex vitro* rooted within 45 days after immersing shoot tips for 24 h in a solution containing 20 mg/l of IBA which saved time and energy for micropropagation of *Strelitzia reginae*.

Keywords: *Strelitzia reginae*, oxidative browning, *ex vitro* rooting

INTRODUCTION

Strelitzia reginae Ait. known as Bird of paradise is a monocotyledon perennial plant originating from South Africa (Figure 1) (Paiva et al., 2004; Chand, 2008). *Strelitzia reginae* is one of the most economically important cut flowers and this is attributed to the exotic characteristics of its colorful flowers, the long length of the stem and the high post-harvest durability (Wood, 1995).

Conventionally *Strelitzia reginae* is propagated either by seed or by division of fans which are naturally developed branches. Both of these propagation methods are very slow and limit its commercial exploitation (Ziv and Halevy 1983; Karnataka 2008). Vegetative propagation by fans division is limited by the low rate of multiplication which is estimated to range from 0.5 to 1.5 divisions per branch and per year. For this type of vegetative propagation plants aged at least 10 years should be used (Vonk Noordegraaf and van der Krogt, 1976; Ziv and Halevy, 1983).

Propagation by seed is is neither a widely used nor a desirable method and even it is not desirable and this is basically due to a several factors including; a prolonged period of juvenility around 4-7 years, seeds dormancy, a great degree of genetic variation in plants developed from seed (Ziv and Halevy, 1983; Van de Pol and Van Hell, 1988; Garcia, 2006). Ziv and Halevy (1983) reported that *Strelitzia reginae* is one of the most important and very scanty cut flower plants for which no uniform cultivars are commercially available for a large scale use.

To overcome those limitations, a reliable propagation method should be developed which may greatly contribute to the selection and the propagation of improved clones of *Strelitzia reginae*. An efficient method by tissue culture could be an excellent and promising alternative to the conventional propagation methods. Chand (2008) noted that despite the economical importance of *Strelitzia reginae*, efficient micropropagation methods are scanty and still not yet developed. The main problem in the *in vitro* propagation of *Strelitzia reginae* is reported to be the accumulation of phenolic compounds specifically quinones on the wounded surface of explants which cause phenolic browning and limit by far their growth and differentiation (North et al., 2010).

The present work aim to investigate the optimal *in vitro* propagation method to cope with oxidative browning and develop a shortened efficient methods that increases the *in vitro* propagation potential and mass production of *Strelitzia reginae* Ait. by combining the steps of rooting and acclimatization.

Figure 1. *Strelitzia reginae* Ait.

Induction of Shoots from Apical Bud Explants

Strelitzia reginae plants were collected from the Higher Institute of Agronomic Sciences, Chott Mariem, Tunisia and utilized as a source of raw material; apical buds were used as explants in this study. The optimal sampling period of *Strelitzia* apical buds was determined based on several previous initiation assays repeated twice each season during a period of two consecutive years. Prior to the establishment of aseptic explants, older leaves of *Strelitzia reginae* were removed. The remaining basal part of the suchers is then cut to leave only apical pieces of a height of 1 cm containing the apical bud surrounded by a few leaves which protect it from disinfecting agents. The raw material was then surface-sterilized with 70% ethanol for a few seconds, and transferred to 10% solution of sodium hypochlorite for 10 min. All traces of the used disinfectant were removed by rinsing several times with sterile distilled water. Before being aseptically excised, surface sterilized explants were immersed in a solution containing 100 mg/l ascorbic acid and 150 mg/l of citric acid to avoid oxidative browning.

The surface sterilized apical bud explants were trimmed down then inoculated on a Murashige and Skoog (1962) basal media (MS) containing (3%) of sucrose and supplemented with different concentrations of activated charcoal (AC) and plant growth regulators: indolebutyric acid (IBA) or indol acetic acid (IAA) and activated charcoal (AC) (Table 1). Media were adjusted to pH 5.8 prior to autoclave at 121°C for 20 minutes. All cultures were maintained for 1month in a growth room with 16h photoperiod at 23 ± 2 °C. Cultures 1, 2, 3 and 4 devoid of charcoal were incubated one week in a total darkness at 23 ± 2 °C. In the present study 24 replicates were used. The new shoots produced were used for the next subsequent experiments.

Table 1. Different concentrations of auxins and activated charcoal on shoot induction of *Strelitzia reginae*

Activated charcoal (g/l)	IAA or IBA (mg/l)			
	0	0,5	1	2
0	1	2	3	4
1	5	6	7	8
2	9	10	11	12
3	13	14	15	16

In the present study *shoot induction* was carried out from apical buds, based on the previous undertaken assays, the optimal sampling period of *Strelitzia* apical buds was determined and was found to be at the end of winter and the beginning of spring with a recovery rate of 75%. The effect of IBA and activated charcoal on shoot induction was also investigated. The obtained results revealed that during the initial stages of culture and regardless of IBA concentrations apical buds maintained on MS media devoid of activated charcoal released brown-black exudates into media. The observed browning is mostly due to the excessive production of polyphenols which led to the death of the inoculated apical buds (Figure 2).

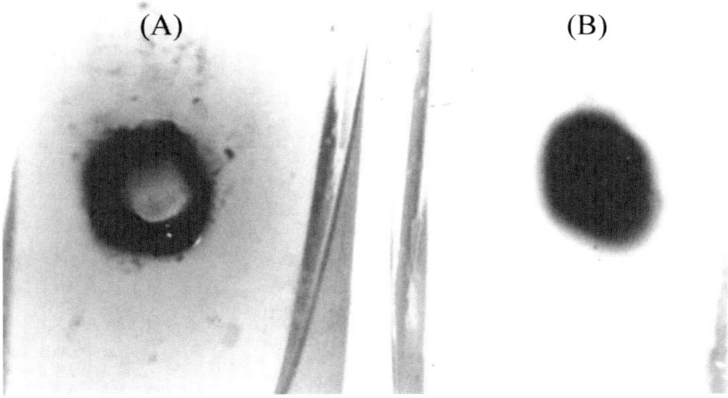

Figure 2. Oxidative browning (A) followed by total necrosis (B) of apical bud of *Strelizia reginae* inoculated on MS medium devoid of activated charcoal.

However, regardless of IBA concentrations apical buds were capable of producing multiple shoots *in vitro* on MS media (13, 14, 15 and 16) supplemented with 3g/l of activated charcoal (Table 1).These results indicated the positive role of activated charcoal in reducing oxidative browning. These findings are in accordance with previous studies that report the high adsorptive ability of activated charcoal to decrease the browning of tissues *in vitro* (Chang et al., 2001; Wang et al., 2005; Guo et al., 2007; Thomas, 2008). Other studies have associated the drastic decrease in phenolic oxidation to the addition of activated charcoal to media (Carlberg et al., 1983; Liu, 1993; Teixeira et al., 1994).

Several reports have shown that the incorporation of activated charcoal in media reduces explants browning by absorbing the inhibitory substances released from wounded explants in Musaceae and Strelitziaceae (Ziv and Halevy, 1983; Diro and van Staden, 2004).

EFFECTS OF AUXINS AND CYTOKININ ON SHOOTS MULTIPLICATION

In vitro shoots of *Strelitzia reginae* of 1 month old obtained from previous assay were split into 3 categories and used for *in vitro* shoots multiplication.

- Shoots with a height below 1 cm
- Shoots with deployed leaf
- Shoots incised at the apex

Shoots were cultured individually in full strength MS medium containing 30 g/l sucrose and supplemented with different concentrations and combinations of IBA and BA at 0, 0.25, 1 and 2 mg/l (Table 2). The medium was adjusted to pH 5.8 and solidified with 6 g/l agar. For each category and each medium, 24 explants were used. A total of 1152 shoots were used to evaluate the effects of auxin and cytokinin by measuring length and number of apical buds after 4 weeks of incubation.

According to the obtained results from multiplication shoots induction in the first study, newly regenerated shoots of *Strelizia reginae* were split into three categories and subcultured continuously on fresh MS medium supplemented with cytokinin and auxin and devoid of activated charcoal.

It is worth noting that most of the shoots showed the presence of necrotic and oxidative browning only one week after inoculation. Results indicated that shoots less than 1 cm in height eventually turn brown and become necrotic. Despite the oxidative browning of a number of shoots, no necrosed shoots with incised apex showed an axillary budding after 4 to 5 weeks of inoculation. In this experiment incised shoots in the apex responded well and performed better than shoots with a height less than 1 cm and no incised apex shoot. There was no positive effect to the addition of IBA alone on axillary budding of inoculated shoots on Medium 5, 9 and 13 (Table 2) and even on apex incised shoots. Inoculated shoots on MS medium supplemented with BA alone at different concentrations exhibited a low multiplication rate (1.2 to 2.2 axillary buds per shoot) with small neoformed axillary buds. Combination of auxin and cytokinin had a positive effect on shoot multiplication. Sturdy and healthy shoots were obtained on MS supplemented 2 mg/l BA and 0.25 mg/l AIB (medium 8) that gave the best multiplication rate (5.4 axillary buds per shoot). It could be concluded that auxin addition improves shoots vigor.

Table 2. Different concentrations of auxins and cytokinin on shoot multiplication of *Strelitzia reginae*

IBA (mg/l)	BA (mg/l)			
	0	0.25	1	2
0	1	2	3	4
0.25	5	6	7	8
1	9	10	11	12
2	13	14	15	16

Three different assays were separately undertaken to attain *ex vitro* rooting from the dissected ends of *in vitro* raised shoots of *Strelitzia reginae* to achieve simultaneous rooting and acclimatization. The first assay aims to evaluate the effect of dipping cut ends of the shoots of 3 to 5 cm of height for 24 h in various concentrations of IBA (0, 2.5, 5, 10, 20 and 40 mg/l) on shoots rooting of *Strelitzia reginae*. For this assay, 48 shoots were used for each concentration. The second assay aims to determine the effect of dipping period of cut ends of the shoots of 3 to 5 cm of height (0, 12, 24 and 48h) in a solution that contains 20 mg/l of IBA on shoots rooting. The third assay aims to optimize the optimal shoot length for these 3 different lengths.

Figure 3. (A) Shoots from apical bud inoculated on initiation medium supplemented with activated charcoal, (B) Axillary shoots inoculated on multiplication medium.

Ex vitro Rooting and Acclimatization of *Strelitzia reginae*

All dipped shoots were transplanted into plastic pots containing peat. The shoots were then allowed to root and to grow for *two 45 days* under greenhouse conditions at a day temperature conditions (25 to 28 °C) and high relative humidity. The first two weeks each pot was watered with half diluted MS Solution basal media and then watered daily with distilled water.

In the present investigation three assays were separately undertaken to optimize *ex vitro* rooting and acclimatization of *Strelitzia reginae* simultaneously. Data presented in table 3 showed that immersion of *Strelizia* shoot tips for 24 h in IBA at different concentrations not only increased rooting percentage but also increased root length and root number per shoot. It appeared that the increase of IBA concentration has been associated with an increase of rooting percentage. The highest rooting percentage (95%) was obtained by dipping shoot tips for 24 h in 20 mg/l of IBA. The highest roots number (2.5 to 3 roots/ shoot) was produced after dipping shoot tips solution at concentrations ranging from 0.25 to 20 mg / l for 24 hours. IBA at a concentration of 10 and 20 mg/l produced the longest roots (> 2 cm). Table 4 shows the effect of immersion of strelitzia shoot tips in a solution of IBA at 20 mg/l for 0, 12, 24, and 48h. Our findings revealed that immersion of shoot tips before transplanting for 24 and 48h strongly and homogeneously improved rooting formation and development. Data presented in table 5 shows that micro-shoots of 3-5 cm length exhibited about 95% of rooting; a maximum of 2.8 roots per shoot with 2 cm root length was produced after immersion of shoot tips for 24 h in 20 mg/l IBA within 45 days. Studies conducted by El Sayid and hegazi (1986) revealed the negative effect of IBA alone or in combination with BA on *in vitro* root development of *Strelitzia reginae*. Bettaieb (2000) mentioned the heterogeneity of *in vitro* rooting of *Strelitzia reginae* by the appearance of pivoting roots and the fasciculated roots. Poor rooting performance was also recorded by Bettaieb (1992) on *Strelitzia* shoots after 30 days of inoculation on MS Medium supplemented with 2 mg/l IBA. The previous undertaken studies highlight the efficiency of *ex vitro* rooting of *in vitro* propagated shoots of *Strelitzia reginae*. The obtained results indicated that immersion of shoot tips for 24 h in a solution containing 20 mg/l of IBA have a positive effect on *ex vitro* rhizogenesis of *Strelitzia reginae* within 45 days by promoting root development compared to control treatment. The rooted shoots under *ex vitro* conditions were well adapted and easy to harden (Figure 3).

Table 3. Effect of IBA concentration on *ex vitro* rooting of *Strelizia reginae*

IBA (mg/l)	0	0.25	5	10	20	40
Rooting percentage (%)	10	37	76	80	95	65
Number of Roots	2.5	3	2.7	2.5	2.7	1.8
Length of Roots (cm)	1.5	1.3	1.5	2.2	2	1.4

Length of shoot = 3 5- cm; Dipping time t = 24h

Table 4. Effect of immersion time of shoot tips of *Strelizia reginae* on *ex vitro* rooting

Dipping time (h)	0	12	24	48
Rooting percentage (%)	10	66.6	95	93
Number of Roots	2.6	2.8	2.8	2.7
Length of Roots (cm)	2	1.8	2	2.1

Length of shoot = 3 5- cm; IBA concentration = 20 mg/l

Table 5. Effect of shoot tips length of *Strelizia reginae* on *ex vitro* rooting

Length of Roots (cm)	**Rooting percentage (%)**	**Number of Roots**	**Length of Roots (cm)**
1 - 3	65	1,7	1,3
3 - 5	95	2,8	2

Dipping time = 24h; IBA concentration = 20 mg/l

Figure 3. (Continued).

Figure 3. Acclimatized Plantlets of *Strelizia reginae*.

CONCLUSION

In conclusion, from the overall results a successful shortened protocol for the micropropagation was established using apical buds and by combining the steps of rooting and acclimatization. This protocol is cost effective and it will contribute to overcome the constraint that of *Strelizia reginae* poses to the horticultural industry worldwide.

REFERENCES

Bettaieb, T. (1992). Generative and *in vitro* multiplication of *Strelizia reginae* Ait. (Oiseau de paradis). G*raduation memory, National Agronomic Institute of Tunisia, 75*.

Bettaieb, T., (2000). Shortening of *Strelizia reginae* Ait. vitroplants production cycle. *Review of the National Agronomic Institute of Tunisia, 15*(2), 19-24.

Carlberg, I., Glimelius, K., and Eriksson, T. (1983). Improved culture ability of potato protoplasts by use of activated charcoal. *Plant Cell Reports,* 2, 223-225.

Chand, M. (2008). *Multiple shoot formation from dissected embryos of Strelitzia reginae (Bird of Paradise).* http://www.urop.uci.edu/symposium/past_symposia/08/program_a-g.pdf.

Chang, S. H., Ho, C. K., Chen, Z. Z., and Tsay, J. Y. (2001). Micropropagation of Taxus mairei from mature trees. *Plant Cell Reports, 20,* 496-502.

Diro, M., and van Staden, J. (2004). Germination of zygotic embryos and *in vitro* growth of seedlings of wild types of Ensete ventricosum. *South African Journal of Botany, 70*(4), 635-639.

Garcia, A. S., Branquinho, E. G. D. A., Menuchi, A. C. T. P., Erlacher, K. C., and Domingues, M. C. S. (2006). *Effect of plant regulators on germination and seed development of Strelitzia reginae*. Thesis São Paulo ano III, *5*, 161-176.

Guo, Y., Bai, J., Zhang, Z. (2007). Plant regeneration from embryogenic suspension-derived protoplasts of ginger (*Zingiber officinale* Rosc.). *Plant Cell Tissue and Organ Culture, 89,* 151-157.

Hegazi, A. E. A. (1986). *Tissue culture propagation of Strelizia reginae Ait. Plant.* Thesis, Faculty of agriculture, Al Azhar University.

Karnataka, J. (2008). Effect of Antioxidant on *In vitro* Establishment of *Strelitzia reginae* Through Shoot Tip Explants. *Journal of Agriculture Science, 21*(2), 324-325.

Liu, M. S. C. (1993). Plant regeneration in cell suspension culture of sugarcane as affected by activated charcoal, medium composition and tissue culture. *Taiwan Sugar.* May-June. p. 18-25.

Murashige, T., and Skoog, F. (1962). A revised medium for rapid growth and bioassays with tobacco cultures. *Physiologia Plantarum, 15,* 473-497.

North, J. J., Ndakidemi, P. A., Laubscher, C. P. (2010). The potential of developing an *in vitro* method for propagating Strelitziaceae. *African Journal of Biotechnology, 9(*45), 7583-7588.

Paiva, P. D. O., Paiva, R., Pasqual, M., and Paiva, L. V. (2004). *In vitro* establishment of *Strelitzia* (*Strelitzia reginae* Banks.) *Ciênc. Agrotecnologia*, *28*(5), 1031-1037.

Teixeira, J. B., Söndahl, M. R., and Kirby, E. G. (1994). Somatic embryogenesis from immature inflorescences of oil palm. *Plant Cell Reports, 13,* 247-250.

Thomas, T. D. (2008). The role of activated charcoal in plant tissue culture. *Biotechnol Advances, 26*, 618-631.

Van de Pol, P. A, and van Hell, T. F. (1988). Vegetative Propagation of *Strelitzia reginae*. *Acta Horticulturae, 226*, 581-586.

Vonk Noordegraaf, C., Van Der Krogt, T. M. (1976). Production and flowering of *Strelitzia*. *Trade Magazine for Floristry*, *31*(49), 18-19.

Wang, Q., Jaana Laamanen, J., Uosukainen, M., Valkonen, J. P. T. (2005). Cryopreservation of *in vitro* grown shoot tips of raspberry (*Rubus idaeus* L.) by encapsulation-vitrification and encapsulation-dehydration. *Plant Cell Reports, 24*, 280-288.

Wood, T. (1995). Ornamental Zingiberaceae. *Revista Brasileira de Horticultura Ornamental, 1,* 12-13.

Ziv, M., and Halevy, A. H. (1983). Control of Oxidative Browning and *in vitro* Propagation of *Strelitzia reginae*. *Horticultural Science, 18*(4), 434-436.

About the Editor

Taoufik Bettaieb is a professor at the National Agronomic Institute of Tunisia, University of Carthage, Tunisia and head of the laboratory of Horticulture sciences. He holds a MSc. and a State Doctorate (Eq. PhD) in Agronomic Sciences from University of Carthage. Currently, he is Visiting Professor jointly at the Higher Institute of Agricultural Sciences of Chott-Mariem. University of Sousse, Tunisia and in the Faculty of Mathematical, Physical and Natural Sciences of Tunis, University of Tunis El Manar. He is also Coordinator of the doctoral committee at the National Agronomic Institute of Tunisia, University of Carthage, Tunisia. He acted as supervisor of 10 PhD projects which 3 under joint sponsorship of Ghent University Belgium, Gembloux Agrobiotech, University of Liège and The National School of Landscape Architecture, Versailles marseilles. He published widely in the field, with more than 90 publications in peer reviewed international journals, 10 chapters and has edited 5 books. Prof. Bettaieb's research is mainly focused on *in vitro* tissue culture, production of cut flowers and foliage, horticultural plants reproduction and breeding techniques of ornamental plants.

INDEX

A

B

C

D

E

G

H

I

L

M

O

P

R

S

T

W

Z